Compact Textbooks in Mathematics

This textbook series presents concise introductions to current topics in mathematics and mainly addresses advanced undergraduates and master students. The concept is to offer small books covering subject matter equivalent to 2- or 3-hour lectures or seminars which are also suitable for self-study. The books provide students and teachers with new perspectives and novel approaches. They may feature examples and exercises to illustrate key concepts and applications of the theoretical contents. The series also includes textbooks specifically speaking to the needs of students from other disciplines such as physics, computer science, engineering, life sciences, finance.

- **compact:** small books presenting the relevant knowledge
- **learning made easy:** examples and exercises illustrate the application of the contents
- **useful for lecturers:** each title can serve as basis and guideline for a semester course/lecture/seminar of 2-3 hours per week.

Gerd Laures • Markus Szymik

A Basic Course in Topology

Gerd Laures
Fakultät für Mathematik
Ruhr-Universität Bochum
Bochum, Germany

Markus Szymik
School of Mathematical
and Physical Sciences
University of Sheffield
Sheffield, UK

ISSN 2296-4568   ISSN 2296-455X (electronic)
Compact Textbooks in Mathematics
ISBN 978-3-662-70601-5   ISBN 978-3-662-70602-2 (eBook)
https://doi.org/10.1007/978-3-662-70602-2

This book is a translation of the original German edition "Grundkurs Topologie," 2nd edition, by Gerd Laures and Markus Szymik, published by Springer-Verlag GmbH, DE in 2015. The translation was done with the help of an artificial intelligence machine translation tool. A subsequent human revision was done primarily in terms of content, so that the book will read stylistically differently from a conventional translation. Springer Nature works continuously to further the development of tools for the production of books and on the related technologies to support the authors.

Translation from the German language edition: "Grundkurs Topologie" by Gerd Laures and Markus Szymik, © Springer-Verlag Berlin Heidelberg 2015. Published by Springer Berlin Heidelberg. All Rights Reserved.

© The Editor(s) (if applicable) and The Author(s), under exclusive license to Springer-Verlag GmbH, DE, part of Springer Nature 2025

This work is subject to copyright. All rights are solely and exclusively licensed by the Publisher, whether the whole or part of the material is concerned, specifically the rights of translation, reprinting, reuse of illustrations, recitation, broadcasting, reproduction on microfilms or in any other physical way, and transmission or information storage and retrieval, electronic adaptation, computer software, or by similar or dissimilar methodology now known or hereafter developed.
The use of general descriptive names, registered names, trademarks, service marks, etc. in this publication does not imply, even in the absence of a specific statement, that such names are exempt from the relevant protective laws and regulations and therefore free for general use.
The publisher, the authors and the editors are safe to assume that the advice and information in this book are believed to be true and accurate at the date of publication. Neither the publisher nor the authors or the editors give a warranty, expressed or implied, with respect to the material contained herein or for any errors or omissions that may have been made. The publisher remains neutral with regard to jurisdictional claims in published maps and institutional affiliations.

This book is published under the imprint Birkhäuser, www.birkhauser-science.com by the registered company Springer-Verlag GmbH, DE
The registered company address is: Heidelberger Platz 3, 14197 Berlin, Germany

If disposing of this product, please recycle the paper.

for Christina and Kirsten

Preface

On Topology Topology, a branch of mathematics, focuses on the qualitative properties of geometric objects. Its conceptual framework is so powerful that nearly all mathematical fields have been influenced by it. As a result, very few areas of modern mathematics are not related—or, as we should say, connected—to topology. Studying topology is, in essence, studying mathematics.

About This Book This book serves as a bridge between introductory lectures on analysis and linear algebra and advanced courses in algebraic and geometric topology. It provides the fundamental topological knowledge that all students of mathematics should acquire during their studies. The book is particularly suitable for students enrolled in bachelor's or master's programmes, especially those whose curricula may only accommodate a one-semester topology course. However, it includes enough material for a full-year course and is ideal for self-study. This makes it a valuable resource for students in mathematical or natural sciences who wish to explore topology in greater depth.

About its History This text is derived from introductory lectures on topology that the authors delivered at Ruhr University Bochum, Germany, during the summer semesters of 2006 and 2008. We focused on using a modern language that unifies the concepts, making them easier to understand.

On Modern Language At the start of their mathematical studies, students are expected to understand the concepts of sets and maps between them. This does not necessarily mean they need to study set theory first; instead, they should master the language of sets to follow lectures effectively. The purpose of category theory is similar. We can view it as a generalisation of set language: categories resemble 'social' sets, where the elements are referred to as objects, and these objects are connected through morphisms. The categorical language helps describe commonly occurring phenomena and constructions by placing them within a unified conceptual framework. This approach makes definitions easier to remember and theorems simpler to express. In summary, learning categorical vocabulary is beneficial and pays off quickly.

On the Content The first five chapters cover the fundamental concepts of point-set topology. Chapters 6 and 7 use paths to explore topological spaces. Together with the theory of coverings presented in Chap. 8, this material forms the core content that most introductory topology courses will address. The final three chapters on bundles, sheaves, and simplicial methods are essential concepts in geometry and topology and serve as a foundation for more advanced courses. Throughout the text, many sections include brief discussions on related topics, offering a starting point for further readings or seminars. The following guide outlines the dependencies among the individual chapters, and each chapter's introduction provides an initial overview of its content.

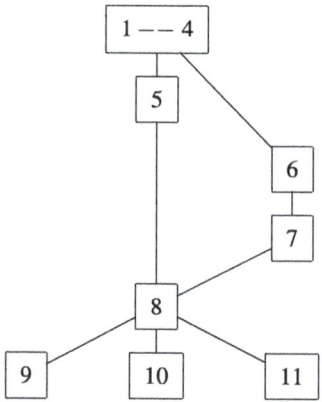

About the Examples and Illustrations The fundamental definitions of topology can be pretty abstract due to its extensive history spanning over a hundred years and its ambition to cover a broad range of applications. This level of abstraction should remain unchanged, and in fact, it is desirable. However, to help students grasp these concepts more quickly, we have supported all definitions with numerous examples. Studying examples is crucial for understanding the theory; without a substantial collection of examples, the theory loses its value. Additionally, we illustrated critical ideas with various figures. While images alone do not provide proof and can sometimes confuse due to their suggestive nature, they can be beneficial in conveying concepts that might be difficult to express with words alone. The dose makes the poison.

About the Exercises The student's creative approach to newly encountered concepts and methods is crucial in acquiring mathematical knowledge. This book, therefore, includes many explicit—and even more implicit—exercises. These are not merely tasks to complete but opportunities for you to engage independently with the material and deepen your understanding. No matter how well-written the book, passive learning through reading is unfortunately ineffective. While it's true that solutions to many problems can be found online, it's essential to approach this differently if you want to maximise your learning. Success comes not from knowing as many solutions as possible but from discovering them independently.

About Schools If there is such a thing as topological schools in Germany, the authors most likely belong to the Heidelberg School. This lineage traces from Threlfall to Seifert and on to Puppe, from whom we have learned extensively, directly and indirectly. The key textbooks from this school include [ST34], [Sch64], [Dol72], [tD91], and [Lüc05]. From the Bonn School associated with Hirzebruch, we should mention [Jän05], [Oss92], and [Kre10]. Additionally, there is a Bochum School, represented by von Querenburg [vQ79] and Stöcker and Zieschang [SZ94]. We have gained much from all these sources. This book distinguishes itself from those previously mentioned by tailoring the selection of material to today's curriculum and utilising categorical language from the outset—reflecting contemporary practices in topology.

About the Translation This text is based on a translation of the third German edition of our textbook [LS23]. A computer produced the initial version. Despite significant technological advances, the output required several rounds of revisions before reaching its current form. We would like to note that the titles of the exercises in the German edition were already quite playful. To maintain—and even enhance—this effect, we have largely kept the titles as they were in the automatic translation.

Acknowledgments We thank Dorothy Mazlum from the publisher for enabling this project and our colleagues James Cranch, Jack Davidson, Andrew Fisher, Oliver House, and Matthew Spong for proofreading the English text.

Bochum, Germany
Sheffield, UK
October 2024

Gerd Laures
Markus Szymik

Contents

1	**Basic Concepts of Topology**		1
	1.1	Metric Spaces	1
	1.2	Topological Spaces	7
	1.3	Closed Subsets	10
	1.4	The Language of Categories	13
2	**Universal Constructions**		19
	2.1	Subspaces	19
	2.2	Products	23
	2.3	Sums	27
	2.4	Identifications and Quotients	30
3	**Connectivity and Separation**		41
	3.1	Connectivity	41
	3.2	Separation and Continuous Extendability	46
4	**Compactness and Mapping Spaces**		55
	4.1	Compactness	55
	4.2	Proper Maps	63
	4.3	Tychonoff's Theorem	68
	4.4	Mapping Spaces	72
	4.5	Locally-Compactly Generated Spaces	78
5	**Transformation Groups**		85
	5.1	Basic Concepts of Equivariant Topology	85
	5.2	Homogeneous Spaces	90
	5.3	Proper Actions	96
6	**Paths and Loops**		101
	6.1	Path Spaces and Loop Spaces	101
	6.2	The Path Component Functor	105
	6.3	The Concept of Homotopy	111
	6.4	Self-maps of the Circle	117

7 Fundamental Groups ... 127
- 7.1 Fundamental Groupoids ... 127
- 7.2 The Seifert–van Kampen Theorem ... 137
- 7.3 Surfaces ... 147

8 Covering Spaces ... 155
- 8.1 The Category of Coverings ... 155
- 8.2 The Lifting Theorem ... 162
- 8.3 Fibre Transport ... 166
- 8.4 The Classification Theorem ... 170
- 8.5 Topological Galois Theory ... 174

9 Bundles and Fibrations ... 183
- 9.1 Fibre Bundles ... 183
- 9.2 Principal Bundles ... 189
- 9.3 Principal Bundles with Discrete Structure Group ... 193
- 9.4 Vector Bundles ... 197
- 9.5 Fibrations ... 199

10 Sheaves ... 207
- 10.1 Presheaves and Sheaves ... 207
- 10.2 Stalks and Étale Spaces ... 210
- 10.3 Sheafification and Pullbacks ... 214

11 Simplicial Sets ... 219
- 11.1 Simplicial Objects and Morphisms ... 219
- 11.2 Singular Simplices and Realisations ... 225
- 11.3 Outlook ... 233

Bibliography ... 239

Index ... 241

Basic Concepts of Topology

Topology is the branch of mathematics dedicated to the study of continuous maps. The concept of a continuous map has already proven important in calculus, for example, when studying maps between metric spaces compatible with the formation of limits. In this chapter, we will start by reviewing this context and then follow on.

1.1 Metric Spaces

We denote by $\mathbb{R}$ the set of real numbers.

Definition 1.1.1

A *metric* on a set X is a map

$$d \colon X \times X \longrightarrow \mathbb{R}$$

that fulfils the following three axioms:

(M1) Positive definiteness: for all points x and y, we have $d(x, y) \geqslant 0$, and $d(x, y) = 0$ holds true if and only if $x = y$.
(M2) Symmetry: we have $d(y, x) = d(x, y)$ for all points x and y.
(M3) Triangle inequality: for all three points x, y, and z, we have

$$d(x, z) \leqslant d(x, y) + d(y, z).$$

The number $d(x, y)$ is called the *distance* from x to y. A *metric space* is a pair (X, d), consisting of a set X and a metric d on X. Usually, instead of the pair (X, d), we only write X.

© The Author(s), under exclusive license to Springer-Verlag GmbH, DE, part of Springer Nature 2025
G. Laures, M. Szymik, *A Basic Course in Topology*, Compact Textbooks in Mathematics, https://doi.org/10.1007/978-3-662-70602-2_1

Examples 1.1.2
The set $\mathbb{R}^n$ of real n–tuples together with the Euclidean metric

$$d(x,y) = \|x-y\|, \quad \text{where} \quad \|x\| = \sqrt{x_1^2 + x_2^2 + \cdots + x_n^2},$$

is a metric space. Any subset X of $\mathbb{R}^n$ becomes a metric space when the distance function d is restricted to X. In this case, we speak of the *induced metric*. Another example of a metric space is any set X together with the *discrete metric*

$$d(x,y) = \begin{cases} 1 & \text{if } x \neq y, \\ 0 & \text{if } x = y. \end{cases}$$

Definition 1.1.3

A map $f\colon X \to Y$ between metric spaces is called *continuous at a point* x in X if, for every $\varepsilon > 0$, there is a $\delta > 0$ such that for all x' in X with $d(x,x') < \delta$ we have

$$d(f(x), f(x')) < \varepsilon.$$

If the map f is continuous at every point x in X, then we say that f is *continuous*.

Examples 1.1.4
Examples of continuous maps between Euclidean spaces should be familiar from analysis. Constant maps are continuous. The identity is continuous. Sums and products of continuous maps are continuous. (Do you recall why?) Polynomials are continuous, and the exponential function is continuous. If the domain X of a map $X \to Y$ carries the discrete metric, the map is always continuous.

Intuitively, continuity means that we can vary the arguments of the map in a small neighbourhood without experiencing too large fluctuations in the values. It is worth changing the definition of a continuous map to be closer to this intuition. Let us do this now:

Definition 1.1.5

Let x be a point in a metric space X and $\varepsilon > 0$. Then the subset of points of X with distance to x smaller than ε is called the *ε–neighbourhood* of x in X. More generally, a *neighbourhood* of x is a subset of X that contains an ε–neighbourhood of x (see Fig. 1.1).

Using these terms, the above definition reads as follows:

▶ **Remark 1.1.6** A map $f\colon X \to Y$ between metric spaces is continuous at x if, for every $\varepsilon > 0$, there is a $\delta > 0$ such that the δ–neighbourhood of x is mapped into the ε–neighbourhood of $f(x)$.

1.1 Metric Spaces

Fig. 1.1 This is not a neighbourhood of x

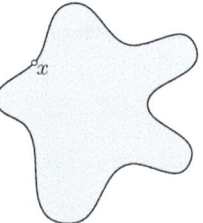

Note that we first choose the ε and then the δ. Therefore, we can also set it up this way:

▶ **Remark 1.1.7** A map $f : X \to Y$ between metric spaces is continuous at x when the pre-image of each ε–neighbourhood V, which is

$$f^{-1}V = \{x \in X \mid f(x) \in V\},$$

contains a δ–neighbourhood of x.

Now, we can even dispense with ε and δ.

Theorem 1.1.8
A map $f : X \to Y$ between metric spaces is continuous at x when the pre-image of each neighbourhood of $f(x)$ is a neighbourhood of x.

Definition 1.1.9

We call a subset of X *open* when it contains, for each of its points, also a δ–neighbourhood of this point.

Examples 1.1.10
The empty set and X itself are always open. The ε–neighbourhood of any point x in X is open, because for each of its elements y, the δ–neighbourhood of y for

$$\delta = \varepsilon - d(x, y)$$

is still entirely in X. This follows from the triangle inequality (see Fig. 1.2).

In an open set, we imagine that there is some space around each point that also lies in the open set. At any rate, we can now say:

Fig. 1.2
The ε–neighbourhoods are open

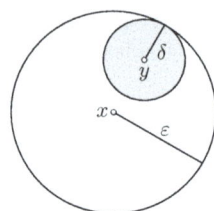

Theorem 1.1.11
A map between metric spaces is continuous if and only if the pre-image of each open set is open.

The characterisation of continuous maps obtained by this theorem can now serve as a starting point for a generalisation. If the set of continuous maps does not depend on the metric but only on the system of open subsets, we should give this structure a name. This generalisation allows the construction of auxiliary spaces that are not metric spaces but help study metric spaces. We would want to take advantage of this idea.

Supplements

Sequential Continuity In metric spaces, continuous maps are those that are compatible with the limit formation of sequences: a sequence (x_n) of elements of the metric space X is called *convergent* to the limit x in X if every neighbourhood of x contains almost all members of the sequence. A map f from X to Y is called *sequentially continuous*, when convergent sequences are mapped to convergent sequences. Continuous maps are always sequentially continuous because the pre-image of a neighbourhood of $f(x)$ is a neighbourhood of x and thus contains almost all sequence members. Conversely, continuity also follows from sequential continuity. If f was not continuous, then there was a neighbourhood V of $f(x)$ whose pre-image U is not a neighbourhood of x. Construct a sequence (x_n) of elements outside of U, where the distance from x_n to x is a null sequence. This sequence converges to x but no value $f(x_n)$ lies in V.

The First Countability Axiom Metric spaces fulfil the *first countability axiom*. It states that for every point x there is a countable set $B(x)$ of neighbourhoods of x with the following property: for every neighbourhood U of x there exists a $V \in B(x)$ with $V \subseteq U$. For example, the set

$$B(x) = \{ U_{1/n}(x) \mid n \in \mathbb{N} \}$$

1.1 Metric Spaces

satisfies the requirements. The existence of these countably many neighbourhoods enables the construction of the sequence in the above proof for the equivalence between continuity and sequential continuity. The first countability axiom also shows the limitations of metric spaces: it can be shown that every system of neighbourhoods on the set of all maps from the unit interval $[0, 1]$ to $\mathbb{R}$ that generates the pointwise convergence of functions violates the axiom (see Sect. 2.2). In particular, there can be no metric for pointwise convergence. There is also a second countability axiom, which we will mention in Sect. 1.2.

Exercises

Exercise 1 Symmetry
Let X be a set. A function $d: X \times X \to \mathbb{R}$ is a metric if and only if

(M1') $d(x, y) = 0 \iff x = y$,
(M2') $d(x, y) \leqslant d(x, z) + d(y, z)$ for all $x, y,$ and z in X.

Does this also apply if (M2') is replaced by the usual triangle inequality

(M3') $d(x, y) \leqslant d(x, z) + d(z, y)$?

Exercise 2 Equivalent Topologies
For points $x = (x_1, x_2)$ of the plane $\mathbb{R}^2$, the norms

$$\|x\|_1 = |x_1| + |x_2|, \quad \|x\|_2 = (x_1^2 + x_2^2)^{1/2}, \quad \|x\|_\infty = \max\{|x_1|, |x_2|\}$$

define metrics d_1, d_2, d_∞ on $\mathbb{R}^2$ by $d_?(x, y) = \|x - y\|_?$. Show that these generate the same concept of convergence.

Exercise 3 Boundedness
Let X be a metric space with the metric d. Show that

$$d'(x, y) = \frac{d(x, y)}{1 + d(x, y)}$$

defines another metric d' that is topologically equivalent to d, i.e., they lead to the same concept of convergence as d.

Exercise 4 French Railway Network
For points x and y of the disk

$$D^2 = \{x \in \mathbb{R}^2 \mid \|x\| \leqslant 1\},$$

let $d(x, y) = \|x - y\|$ if x and y are on the same line through the origin, otherwise $d(x, y) = \|x\| + \|y\|$. Show that d is a metric and that it induces the discrete metric on the subspace

$$S^1 = \{x \in \mathbb{R}^2 \mid \|x\| = 1\}$$

What do the neighbourhoods of $(0, 0)$ and $(1/2, 0)$ look like?

Exercise 5 Lengths

As is well known, every permutation f of $\{1, \ldots, n\}$ can be written as a product of transpositions of adjacent elements, i.e.,

$$f = (a_1, a_1 + 1) \circ \cdots \circ (a_k, a_k + 1)$$

with $k \geqslant 0$ and $a_j \in \{1, \ldots, n-1\}$ for all $j \in \{1, \ldots, k\}$. However, such a presentation is not unique. The *length* $L(f)$ of f is the minimal k for which there is a presentation as above. Calculate the lengths of all permutations of $\{1, 2, 3\}$. Show that

$$d_L(f, g) = L(f^{-1} \circ g)$$

defines a metric d_L on the set of permutations of $\{1, \ldots, n\}$.

Exercise 6 Fixed Points

For a permutation f of $\{1, \ldots, n\}$ be $M(f)$ the number of non-fixed points of f. Show that

$$d(f, g) = M(f^{-1} \circ g)$$

defines a metric d on the set of permutations of $\{1, ..., n\}$.

Exercise 7 Evaluations

Let p be a prime number. For an integer $a \neq 0$, set

$$v_p(a) = \max\{n \in \mathbb{N} \mid p^n \text{ divides } a\}.$$

For integers x and y, set

$$d_p(x, y) = \begin{cases} p^{-v_p(x-y)} & x \neq y, \\ 0 & x = y. \end{cases}$$

Show that d_p is a metric on $\mathbb{Z}$. The completion of $\mathbb{Z}$ with respect to this metric is again a ring, the ring $\mathbb{Z}_p$ of the *p*–adic integers. In this context, the number $v_p(a)$ is called the *p–adic evaluation* of a.

Exercise 8 Identity and Evaluation

Let F be the set of all continuous functions $[0, 1] \to \mathbb{R}$. Define metrics on F by

$$d_\infty(f, g) = \sup_x |f(x) - g(x)|,$$

$$d_2(f, g) = \left(\int_0^1 (f(x) - g(x))^2 \, dx \right)^{1/2}.$$

Investigate which of the following maps are continuous:

(a) $\text{id}: (F, d_\infty) \longrightarrow (F, d_2)$
(b) $\text{id}: (F, d_2) \longrightarrow (F, d_\infty)$
(c) $\text{ev}_0: (F, d_\infty) \longrightarrow \mathbb{R}, \ f \longmapsto f(0)$
(d) $\text{ev}_0: (F, d_2) \longrightarrow \mathbb{R}, \ f \longmapsto f(0)$

1.2 Topological Spaces

Exercise 9 Points of Discontinuity
Provide examples of functions $f: \mathbb{R} \to \mathbb{R}$ that are continuous at the following points:

(a) nowhere,
(b) on the set of all irrational numbers $\mathbb{R} \setminus \mathbb{Q}$ and otherwise not,
(c) at the point 0 and otherwise not.

Is there a function that is continuous on $\mathbb{Q}$ and otherwise not?

Exercise 10 The Small One Plus One
Is there a map $f: \mathbb{R} \to \mathbb{R}$ that satisfies the conditions $f(x+y) = f(x) + f(y)$ and $f(x \cdot y) = f(x) \cdot f(y)$ for all real numbers x and y and that is not continuous?

1.2 Topological Spaces

As soon as we recognise that the continuity of maps between metric spaces can be characterised by using open sets, it is easy to talk about continuous maps $f: X \to Y$ even if X and Y are not necessarily metric spaces: it is sufficient if specific subsets of X and Y are distinguished as being open. It has proven helpful to require that the set of open subsets has some additional properties.

Definition 1.2.1

A *topological structure* or briefly a *topology* on a set X is a set $\mathcal{T}$ of subsets of X, called the *open* subsets, such that the following three axioms hold.

(T1) The subsets $\emptyset$ and X are open.
(T2) The intersection of finitely many open subsets is open.
(T3) The union of any number of open subsets is open.

A *topological space* is a pair $(X, \mathcal{T})$ consisting of a set X and a topological structure $\mathcal{T}$ on X. We usually refer to a topological space $(X, \mathcal{T})$ simply as X. A subset M of a topological space X is called a *neighbourhood of $x \in M$* if there is an open set U that satisfies $x \in U \subseteq M$.

From (T3), we find:

▶ **Remark 1.2.2** The open subsets of a topological space are those that are neighbourhoods of each of their points.

Example 1.2.3
There are many ways to provide a set with a topology. For example, Fig. 1.3 shows all possible topologies (without permutations of the elements) of a three-element set.

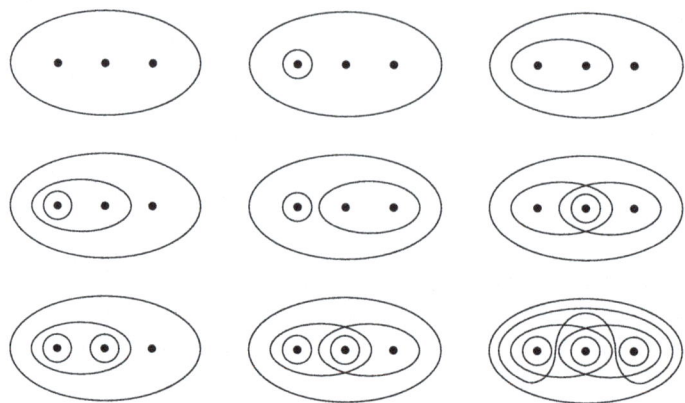

Fig. 1.3 The different topologies on a three-element set

Example 1.2.4
Let (X, d) be a metric space. Then the set $\mathcal{T}(d)$ of open sets is a topological structure on X. It is perhaps already clear why we do not also require that the intersections of any set of open sets should again be open. This is not correct in metric spaces. Although $\{x\}$ is the intersection of the ε-neighbourhoods of x, the subset $\{x\}$ is not necessarily open. For $X = \mathbb{R}$ with the standard metric, this is not the case. Furthermore, note that there can be different metrics $d \neq d'$ on the same set X so that the corresponding topologies coincide, i.e., we can have $\mathcal{T}(d) = \mathcal{T}(d')$ even if $d \neq d'$ (see the exercises from the preceding section).

Example 1.2.5
If X is any set, the power set of X defines a topology on X. It is called the *discrete topology* and X is called a discrete space. This topology coincides with the topology that comes from the discrete metric.

Example 1.2.6
The *clump topology* (or *indiscrete topology*) on a set consists only of the empty set and the set itself.

The topologies on a set X are ordered by inclusion. We say that a topology $\mathcal{T}_1$ is *coarser* than a topology $\mathcal{T}_2$ (and $\mathcal{T}_2$ is then correspondingly *finer* than $\mathcal{T}_1$) if $\mathcal{T}_1 \subseteq \mathcal{T}_2$ holds, i.e. every open set of $(X, \mathcal{T}_1)$ is also open in $(X, \mathcal{T}_2)$. This is the case if and only if the identity map

$$(X, \mathcal{T}_2) \longrightarrow (X, \mathcal{T}_1)$$

is continuous. The clump topology is the coarsest topology on X; the discrete topology is the finest. Examples in between—and these are the most exciting topologies—are given by the metric spaces. More will follow in the exercises and in the course of this book.

Definition 1.2.7

A map $f: X \to Y$ between topological spaces is called *continuous* at x if the pre-images of the neighbourhoods of $f(x)$ are neighbourhoods of x. A map between topological spaces is called *continuous*, if the pre-images of open sets are open or, equivalently, if f is continuous at every point.

This definition was motivated in the previous section on metric spaces, and it is only repeated here for clarity. It should never be forgotten that it is the continuous maps that are studied in topology; the topological spaces appear only as their supporting structure. When first entering the field of topology, the concept of a continuous map is often already provided with intuition from calculus courses. At the same time, topological spaces are still unfamiliar in their generality. Therefore, in the following sections, topological spaces will be somewhat in the foreground before gradually realising that they can only be understood with the help of continuous maps.

Supplements

Metrisability A topological space $(X, \mathcal{T})$ is called *metrisable* if there is a metric d on X such that $\mathcal{T} = \mathcal{T}(d)$. We have seen in a supplement of the preceding section that the first countability axiom is necessary for this. A sufficient criterion is the *second countability axiom*, which states that the countable set $B(x)$ can be chosen independently of x (at least for spaces that fulfil the third separation axiom, see Sect. 3.2). However, this is not necessary. A characterisation was found by Bing, Nagata and Smirnov, see 10.B in [vQ79].

Convergence The convergence of sequences in topological spaces is defined in the same way as in metric spaces, through neighbourhoods. However, in general topological spaces, unlike in metric spaces, there can be sequences that converge to two different limits. For example, if we equip the set $\{0, 1\}$ with the clump topology, any sequence converges to 0 and to 1. The limit value is unique if X satisfies the Hausdorff property, i.e., any two points can be separated by neighbourhoods (see Sect. 3.2).

Exercises

Exercise 11 Finite-Complement Topology
For any set X, show that the subsets of X whose complements are finite or all of X form a topology on X.

Exercise 12 Order Topology
Let X be a linearly ordered set. (This can be looked up in Sect. 1.4.) We write $a < x$ if $a \leqslant x$ and $a \neq x$. Consider subsets of the form $\{x \in X \mid a < x < b\}$ as well as the two subsets $\{x \in X \mid$

$k < x\}$ and $\{x \in X \mid x < g\}$, if there is a smallest element k or a largest element g. Show that the set of unions of such sets is a topology on X.

Exercise 13 Set Theory
Let $f\colon X \to X'$ be a map between sets X and X'. Examine whether the following statements hold in general. If not, provide as simple as possible conditions for f that are necessary and sufficient.

(a) $f(A \cup B) = f(A) \cup f(B)$ (a') $f^{-1}(A' \cup B') = f^{-1}(A') \cup f^{-1}(B')$
(b) $f(A \cap B) = f(A) \cap f(B)$ (b') $f^{-1}(A' \cap B') = f^{-1}(A') \cap f^{-1}(B')$
(c) $f(X \setminus A) = X' \setminus f(A)$ (c') $f^{-1}(X' \setminus A') = X \setminus f^{-1}(A')$
(d) $f^{-1}(f(A)) \subseteq A$ (d') $f(f^{-1}(A')) \subseteq A'$
(e) $f^{-1}(f(A)) \supseteq A$ (e') $f(f^{-1}(A')) \supseteq A'$

for all $A, B \subseteq X$ or all $A', B' \subseteq X'$.

Exercise 14 The Zariski Topology
Let A be a commutative ring with unit. Let $\mathrm{Spec}(A)$ be the set of prime ideals of A. A *root* of an element f of A is a prime ideal P such that we have $f/P = 0$ in the residue ring A/P, that is, when the element f is contained in P. For each f in A, let $N(f) \subseteq \mathrm{Spec}(A)$ be the set of roots of f. For each subset S of A, let $N(S) = \bigcap_{f \in S} N(f)$ and $U(S)$ be its complement. Show that

$$\{U(S) \subseteq \mathrm{Spec}(A) \mid S \subseteq A\}$$

is a topological structure on $\mathrm{Spec}(A)$. Such topologies play a fundamental role in algebraic geometry.

1.3 Closed Subsets

Definition 1.3.1

A subset A of a topological space X is called *closed in X*, if its complement $X \setminus A$ is open in X.

Thus, the meaning of 'closed' is not 'the opposite of open'. A subset can be both open in X and closed in X, such as $\emptyset$ and X in any space X. And some subsets are neither open nor closed, such as $\mathbb{Q}$ in $\mathbb{R}$. Here, avoidable beginner mistakes lurk.

▶ **Remark 1.3.2** Arbitrary intersections and finite unions of closed sets are closed.

This follows directly from *De Morgan's Laws:*

The complement of the union is the intersection of the complements.
The complement of the intersection is the union of the complements.

In formulas:

$$X \setminus \Big(\bigcup_{i \in I} U_i\Big) = \bigcap_{i \in I} (X \setminus U_i)$$

1.3 Closed Subsets

$$X \setminus \left(\bigcap_{i \in I} U_i\right) = \bigcup_{i \in I} (X \setminus U_i)$$

In addition, we have the following statement.

▶ **Remark 1.3.3** A map $f: X \to Y$ between topological spaces X and Y is continuous if and only if the pre-images of the closed subsets of Y are all closed in X.

Therefore, we could also have built the set-theoretic topology on the concept of closed subsets instead of open subsets.

Definition 1.3.4

The intersection of all closed sets that contain M is again a closed subset of X that contains M. It is called the *closure* of M in X and is denoted by $\overline{M}$. A subset M of a topological space is called *dense* in X if we have $\overline{M} = X$.

Examples 1.3.5
Intuitively, the closure of M corresponds to the set of all points that touch the set M (see also the following supplement). For example, the closure of the subset $M = \{1/n \mid n \in \mathbb{N}\}$ of $\mathbb{R}$ is the set $M \cup \{0\}$. Every closed set A must contain the zero point with M because otherwise $\mathbb{R} \setminus A$ is an open neighbourhood of 0 that does not intersect with M. Similarly, we can argue that the closure of $[0, 1[$ in $\mathbb{R}$ is the interval $[0, 1]$ or that $\mathbb{Q}$ is dense in $\mathbb{R}$.

▶ **Remark 1.3.6** A subset M is closed if and only if we have $\overline{M} = M$. In particular, we always have $\overline{\overline{M}} = \overline{M}$.

Definition 1.3.7

Let M be a subset of a topological space X. The set

$$\overset{\circ}{M} = X \setminus (\overline{X \setminus M})$$

is called the *interior* of M. Its complement

$$\partial M = \overline{M} \setminus \overset{\circ}{M}$$

is called the *boundary* of the set M in X.

▶ **Remark 1.3.8** The set $\overset{\circ}{M}$ is the largest open set that is contained in M.

Examples 1.3.9
The boundary of $\mathbb{Q}$ in $\mathbb{R}$ is $\mathbb{R}$. The boundary of a non-degenerate interval is the set of its endpoints.

Definition 1.3.10

A continuous map $f\colon X \to Y$ between topological spaces X and Y is called *closed* (respectively, *open*) if the image of every closed (respectively, open) set of X is closed (respectively, open) in Y.

Examples 1.3.11
The map
$$f\colon \mathbb{R} \longrightarrow \mathbb{R},\ x \longmapsto x$$
is both closed and open. The map
$$f\colon \mathbb{R} \longrightarrow \mathbb{R},\ x \longmapsto 0$$
is closed but not open. The map
$$f\colon \mathbb{R} \longrightarrow \mathbb{R},\ x \longmapsto \arctan(x)$$
is not closed but open. The map
$$f\colon \mathbb{R} \longrightarrow \mathbb{R},\ x \longmapsto |\arctan(x)|$$
is neither closed nor open.

Supplement

Accumulation Points A point x of a topological space X is called an *accumulation point* of $M \subseteq X$ if every neighbourhood of x intersects the set M (see Fig. 1.4). The set of accumulation points contains M and it is closed. This can be seen as follows: assume that x has a neighbourhood U that does not intersect M. Without loss of generality, we may assume that U is open. The complement of U is therefore closed and contains M. So it also contains $\overline{M}$. Since U does not intersect the closure, we may conclude that the complement of the accumulation points is open. Elements in the closure are, therefore, always accumulation points. The converse also applies: every closed set A that contains M also contains the accumulation points because the complement of A is an open neighbourhood that does not intersect M.

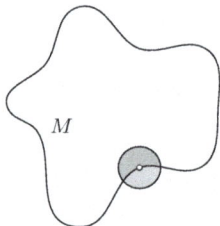

Fig. 1.4 This is an accumulation point of M

Exercises

Exercise 15 Boundaries
Determine the boundaries of the following subsets of $\mathbb{R}^2$:

(a) $M = \{(x, y) \mid x > 0 \text{ and } y \neq 0\}$
(b) $M = \{(x, y) \mid 0 \leqslant x^2 - y^2 < 1\}$

Exercise 16 Gluing of Continuous Functions
Let $X = A \cup B$, where A and B are closed. Let $f: A \to Y$ and $g: B \to Y$ be continuous and $f(x) = g(x)$ for all x in the intersection $A \cap B$. Show that the map $h: X \to Y$ with $h(x) = f(x)$ for $x \in A$ and $h(x) = g(x)$ for $x \in B$ is well-defined and continuous.

Exercise 17 Three Times Is Once
Suppose A is a closed subset of a topological space X. Then so is $A' = \overline{X \setminus A}$. Show that we have $A''' = A'$.

Exercise 18 Sport
How many sets (at most) can be formed from a subset M of $\mathbb{R}$ by closure and complement formation?

Exercise 19 Kuratowski's Closure Axioms
Let X be a set and h a map of the power set of X into itself with the following properties:

(K1) $h(\emptyset) = \emptyset$
(K2) $A \subseteq h(A)$
(K3) $h(h(A)) = h(A)$
(K4) $h(A \cup B) = h(A) \cup h(B)$

for all $A, B \subseteq X$. Show that there is a unique topology on X so that for each subset A in X, the set $h(A)$ is the closure of A with respect to this topology.

Exercise 20 Closure
Show that a map f between topological spaces is continuous if and only if $f(\overline{M})$ is contained in $\overline{f(M)}$ for all subsets M of the source.

1.4 The Language of Categories

As already emphasised, the class of topological spaces becomes interesting only because we can consider the set of continuous maps between any two topological spaces. The following remark results directly from the definition of continuity. Its significance is enormous, enabling us to compose simple continuous maps to construct more complicated ones.

▶ **Remark 1.4.1** If X is a topological space, then the identity $\mathrm{id}_X: X \to X$ is a continuous map. If $f: X \to Y$ and $g: Y \to Z$ are continuous maps, then also the composition $gf: X \to Z$ is continuous (see Fig. 1.5).

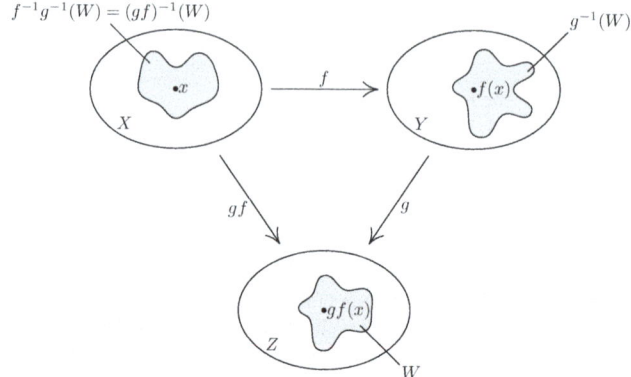

Fig. 1.5 The composition of continuous maps is continuous

In more sophisticated terminology, but also much shorter, we can rephrase the remark to say that the class of topological spaces and the continuous maps between them form a category. This section will explain what that means. The language of categories is well suited to capturing frequently recurring phenomena and constructions across mathematics in a unified conceptual framework. Learning the new vocabulary will quickly pay off. The standard reference is [Mac98].

Definition 1.4.2

A *category* $\mathcal{C}$ consists of the following data. First, a class whose elements are called *objects*. Then for every two objects X and Y, a set $\text{Mor}_{\mathcal{C}}(X, Y)$ whose elements are called *morphisms*. Instead of $f \in \text{Mor}_{\mathcal{C}}(X, Y)$, we often write $f \colon X \to Y$. For every three objects X, Y and Z, we need a map

$$\text{Mor}_{\mathcal{C}}(Y, Z) \times \text{Mor}_{\mathcal{C}}(X, Y) \to \text{Mor}_{\mathcal{C}}(X, Z), \quad (g, f) \mapsto gf,$$

called *composition*. Finally, for each object X, there must be a distinguished element id_X in the set $\text{Mor}_{\mathcal{C}}(X, X)$, the *identity* of X. The only axioms that these data should satisfy are the associativity of the composition, which means

$$h(gf) = (hg)f$$

whenever the compositions are defined, and the neutrality of the identities, which means

$$f \, \text{id}_X = f = \text{id}_Y \, f$$

for all morphisms $f \colon X \to Y$.

1.4 The Language of Categories

Before giving the first examples of categories, the word 'class' should be commented on. We used it in the definition because the objects of many categories do not form a set in the traditional sense; we want to avoid the famous contradictions of set theory. Thus, we can speak of a class whose objects are all sets but not of the set of all sets. Categories whose objects form a set are called *small*. Without any bad conscience about this detail, we recommend moving on and immediately focusing on the interesting examples.

Examples 1.4.3
There are plenty of examples of categories. In many examples of categories, the objects are sets 'with structure' and the morphisms are the 'structure-preserving' maps. For example, there is the category **Sets** of sets and maps, the category **Grp** of groups and group homomorphisms, the category **AbGrp** of abelian groups and their homomorphisms, and the category of rings and ring homomorphisms. If K is a field, there is the category of K–vector spaces and K–linear maps. In short, algebra is full of categories. Topology begins by defining the category **Top** of topological spaces and continuous maps. Algebraic topology deals, among other things, with mapping these (or similar categories of topological objects) into categories of algebraic objects to tackle topological problems with algebraic help. The maps between categories have their own name: functors. We will explain those later when we need them (see Sect. 6.2).

From every category $\mathcal{C}$, the opposite category $\mathcal{C}^{op}$ can be produced by reversing the arrow directions. Both categories have the same objects, but the morphisms $X \to Y$ in $\mathcal{C}^{op}$ are given by the morphisms $Y \to X$ in $\mathcal{C}$. At first glance, this does not seem very interesting, but it is very useful for theoretical purposes. There is a so-called 'dual' concept for every word of the category language. It is obtained by reversing all arrows. One concept then often differs from the other only by the prefix 'co-'. We will soon get to know examples: products and coproducts, fibrations and cofibrations, simplicial and cosimplicial...

Definition 1.4.4

A morphism $f \colon X \to Y$ of a category $\mathcal{C}$ is called an *isomorphism* if there is a morphism $g \colon Y \to X$ in the opposite direction so that $gf = \mathrm{id}_X$ and $fg = \mathrm{id}_Y$ hold. (We can show that such an inverse, if it exists, is always unique.) The isomorphisms in the category of topological spaces and continuous maps are, by the way, called *homeomorphisms*.

Two topological spaces X and Y are, therefore, homeomorphic if there are continuous maps $f \colon X \to Y$ and $g \colon Y \to X$ so that $gf = \mathrm{id}_X$ and $fg = \mathrm{id}_Y$ hold. Two homeomorphic spaces are considered equivalent in topology, and one of the fundamental problems of topology is figuring out whether two given spaces are homeomorphic or not. If they turn out to be homeomorphic, the question immediately arises how many homeomorphisms there are between them.

A critical warning right at this point is that homeomorphisms are automatically bijective, but not every continuous bijection is a homeomorphism. For example, we can provide each set with the discrete topology and with the clump topology. The identity is a continuous map from the discrete topology into the clump topology. As soon as the set has at least two different elements, the inverse map is not continuous. There are, however, classes of topological spaces between which continuous bijections are already homeomorphisms. A corresponding theorem can be found in Sect. 4.1.

Definition 1.4.5

Given an object X in a category, a morphism $f\colon X \to X$ is called an *endomorphism* of X. It is called an *automorphism* of X if it is an isomorphism. The automorphisms of any object form a group with respect to the composition, with the identity as a neutral element, the *automorphism group* $\mathrm{Aut}_\mathcal{C}(X)$.

Many important groups appear as automorphism groups: the symmetric groups are the automorphism groups of the set $\{1, \ldots, n\}$, and the automorphism groups of the K-vector spaces K^n are the general linear groups $\mathrm{GL}(n, K)$.

Example 1.4.6

Every group G appears as an automorphism group of an object of a category. For example, we can consider the category with a unique object whose (only) set of morphisms is G. The composition and identity are then determined by the given group structure. This is then a small category because it only has one object. The automorphism group of this object is the group G. Therefore, this category itself is also denoted by G. Groups are essentially the same as small categories with a unique object whose endomorphisms are isomorphisms.

If a composition

$$X \xrightarrow{s} Y \xrightarrow{r} X$$

is the identity of X, so that we have $rs = \mathrm{id}_X$, then s is called a *section* (or *right inverse*) of r and r a *retraction* (or *left inverse*) of s. We then also call X a *retract* of Y.

Supplement

Partially Ordered Sets A *partially ordered set* is a small category in which each set of morphisms has at most one element, and in which every isomorphism is an identity. We write $X \leqslant Y$ if there is an arrow $X \to Y$. A partially ordered set is *linearly ordered* if there is a unique morphism between any two elements. If $(X, \mathcal{T}_X)$ is a topological space then the topology $\mathcal{T}_X$ is a category via the inclusions of the open subsets among each other. This is a partially ordered set that is not linearly ordered in general. These categories will play a major role in Chap. 10. The partially ordered set $\{0 \leqslant 1 \leqslant 2 \leqslant \cdots \leqslant n\}$ is denoted by $[n]$. It is linearly ordered. The objects of $[n]$ are the $n+1$ numbers $0, \ldots, n$. Thus, the number n does not stand for the number of objects but for the 'dimension' of the category: we imagine the

1.4 The Language of Categories

objects of $[n]$ as the corners of an n–simplex (see Chap. 11, where these categories will play a major role).

Exercises

Exercise 21 Right and Left Inverses
Let f, g, and h be morphisms in a category $\mathcal{C}$ for which $gf = \mathrm{id}$ and $fh = \mathrm{id}$ hold. Show that f is an isomorphism and that $g = h$ holds.

Exercise 22 Homeomorphism
How many pairwise non-homeomorphic topological spaces with two elements are there? The precise determination of the number of homeomorphism classes of finite spaces with a predetermined number of elements is a problem that still needs to be solved. See, for example, [Ern74] and also [Sto66] for more on finite topological spaces.

Exercise 23 Three Times Is not a Charm
Let $\mathcal{T}$ be a topology on the three-element set $X = \{1, 2, 3\}$. Then the homeomorphism group of $(X, \mathcal{T})$ is a subgroup of the symmetric group with $3! = 6$ elements. Show that there is no topology $\mathcal{T}$ on this set X whose homeomorphism group has exactly three elements. Is there even a topological space whose homomorphism group has exactly three elements?

Universal Constructions

2

The first chapter introduced the category of topological spaces and continuous maps. This chapter will describe constructions that allow us to create new topological spaces. Knowing the points and open sets of the new spaces is often less important. Much more important is to understand how the new spaces are related to the old ones, i.e., which continuous maps exist into the new spaces or out of them. We will describe each case using a so-called universal property of the new spaces. Experience shows that beginners often find it difficult to ignore the points and open sets to work with the universal properties instead. However, the latter is essential in topology (and also in many other areas of mathematics) and, therefore, needs to be practised as early as possible.

2.1 Subspaces

Let X be a topological space, let M be a set and

$$f: M \longrightarrow X$$

a map. Then we would like to have a topological structure on M, and not just any, but one that has something to do with the topology on X and with f. This section clarifies which requirements are reasonable on such a topology and shows that these can be fulfilled in precisely one way.

The topology on M should make the map f continuous. This is the case when for each open subset U of X, the subset $f^{-1}U$ is open in M. At least these sets must, therefore, be open in M. Despite this requirement, the topological structure is not yet fixed; we could declare more subsets than these open, and f would still be continuous. For example, every map is continuous for the discrete topology on M. To distinguish a topological structure on M, we must, therefore, impose another

Fig. 2.1 The subspace topology

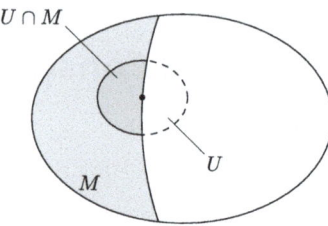

condition to guarantee that there are not too many open sets in M. It is easy to see that the set

$$\mathcal{T} = \{\, f^{-1}U \mid U \text{ is open in } X \,\}$$

of subsets already forms a topological structure on M. According to what has just been said, it is the coarsest topology on M, for which f is continuous.

Definition 2.1.1

The topology $\mathcal{T}$ is called the *induced topology*. If f is the inclusion of a subset $M \subseteq X$, i.e., we have

$$f : M \longrightarrow X, \ x \mapsto x,$$

then the topology $\mathcal{T}$ is called the *subspace topology*. In this case, we can write $\mathcal{T}$ as follows (see Fig. 2.1):

$$\mathcal{T} = \{\, U \cap M \mid U \text{ is open in } X \,\}.$$

Example 2.1.2
In the subspace $M = [\,0, 1\,] \cup \{2\}$ of $\mathbb{R}$, the subset $\{2\}$ is open, even though it is not open in $\mathbb{R}$. Namely, we have

$$\{2\} = M \cap\,]1, 3[,$$

and the interval $]1, 3[$ is open in $\mathbb{R}$. Similarly, we see that $]0, 1]$ is also open in M. By the way, this subspace topology on M corresponds to the topology of the induced metric. This applies to every subspace M of a metric space X because the ε–neighbourhoods of M are the intersections of the ε–neighbourhoods of X with M. The induced metric generates the induced topology because we can write every open set as a union of ε–neighbourhoods.

We can characterise the induced topology by the sets of all continuous maps to M. Here is how it works. Clearly, if $f : M \to X$ is continuous, then for any continuous map $g : T \to M$, the composition $fg : T \to X$ is also continuous.

2.1 Subspaces

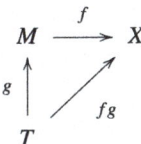

The letter 'T' stands for *test space*; the map g from T to M tests M. The following theorem states that there is a unique topological structure on M that also fulfils the converse property.

> **Theorem 2.1.3 (Universal Property of the Induced Topology)**
> *The topology induced by $f: M \to X$ is the only topology on M with the property that any map $g: T \to M$ is continuous if and only if the composition $fg: T \to X$ is continuous.*

Proof. Let us assume that M carries the induced topology and fg is continuous. Let U' be an open subset of M. Then U' is the pre-image of an open set U under f. Therefore, the pre-image of U' under g is also open because
$$g^{-1}U' = g^{-1}f^{-1}U = (fg)^{-1}U.$$
It follows that the induced topology fulfils the property.

The uniqueness of the topology can be seen by applying the condition to the map g that is the identity on M for potentially different topologies. In more detail, let $\mathfrak{J}$ and $\mathfrak{J}'$ be two topologies on M that meet the requirement stated in the theorem. Then f is continuous for each of the two topologies on M because the identity map between identical topological spaces is always continuous. The diagram

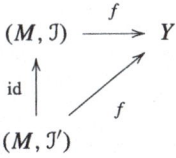

shows the identity on M from $\mathfrak{J}'$ to $\mathfrak{J}$ is continuous. It follows that we have the inclusion $\mathfrak{J} \subseteq \mathfrak{J}'$. When we swap the roles of $\mathfrak{J}$ and $\mathfrak{J}'$, we see that the opposite inclusion also holds, proving the topologies' equality. □

A property that characterises an object through the morphisms into or out of it is often called a *universal property*. The above theorem thus gives the universal property of the induced topology. Further examples of universal properties will follow. Unless otherwise written, subsets of topological spaces will, from now on, always be provided with the subspace topology. Again, we emphasise that 'open' and 'closed' are relative terms; every subspace X of a topological space Y is open and closed in X, but not necessarily in Y. Finally, it should be noted, to counteract confusion, that in vector spaces that carry a topology, in $\mathbb{R}^n$ for example, every subvector space is, of course, a subspace in the sense of topology, but not every subspace also needs to be a vector subspace. Interesting subspaces that are not vector subspaces are, for example, the spheres

$$S^n = \{x \in \mathbb{R}^{n+1} \mid \|x\| = 1\}.$$

Topology is far from being able to answer all questions about continuous maps between spheres; however, there are other goals.

Definition 2.1.4

A continuous map $f \colon X \to Y$ is an *embedding*, if f is injective and X carries the topology induced by f.

▶ **Remark 2.1.5** A continuous map $f \colon X \to Y$ is an embedding if and only if it is a homeomorphism onto its image. In this case, the image carries the subspace topology.

Example 2.1.6
A *knot* is a circle S^1 embedded in $\mathbb{R}^3$. The trefoil knot (see Fig. 2.2) and the figure-eight knot (see Fig. 2.3) are shown below.
In knot theory, we investigate when two such knots can continuously be transformed into each other without the knots being cut. This is obviously a topological question. A readable introduction to classical knot theory is provided by Burde and Zieschang [BZ03].

Fig. 2.2 The trefoil knot

Fig. 2.3 The figure-eight knot

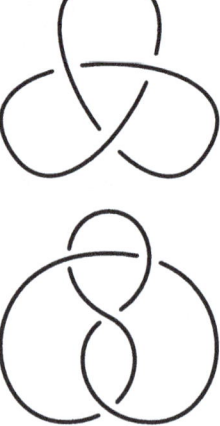

Exercises

Exercise 24 A Consistency Test
Let X be a subset of a metric space Y. Then X is also a metric space by restriction of the metric. The corresponding topology on X is the subspace topology. In particular, subspaces of metrisable spaces are themselves metrisable.

Exercise 25 Induction Is Transitive
Let X be a topological space. Specify and prove that every subspace of a subspace of X is also a subspace of X.

Exercise 26 Subspace Description
Characterise the subspace topology through (a) closed sets, (b) the closure operator, and (c) neighbourhoods.

Exercise 27 A Trip to Babylon
Let X be a subspace of a topological space Y. The inclusion is open if and only if X is open in Y. Does this also hold for 'closed' instead of 'open'?

Exercise 28 The Topology Induced by the Parabola
Describe the open sets of the topology on $\mathbb{R}$ that is induced by the map $\mathbb{R} \to \mathbb{R}$, $x \mapsto x^2$. Is there a metric on $\mathbb{R}$ that generates the same topology?

Exercise 29 Alphabet
Consider the letters of the alphabet

$$\text{A B C D E F G H I J K L M N O P Q R S T U V W X Y Z}$$

as subspaces of $\mathbb{R}^2$ and find the sets of letters that are homeomorphic to each other.

2.2 Products

In this section, the situation from the last section is generalised from one map $f : M \to X$ to several maps. We assume that we are given a set M, topological spaces X_i for an index set I, and maps

$$f_i : M \longrightarrow X_i$$

for each $i \in I$. Again, we seek a topology $\mathcal{T}$ on M that is as coarse as possible so that all f_i are continuous. For this to happen, the set

$$\mathcal{S} = \bigcup_{i \in I} \{ f_i^{-1}(U) \mid U \text{ is open in } X_i \}$$

of subsets of M must be contained in $\mathcal{T}$. Unfortunately, this set $\mathcal{S}$ does not generally form a topology because $\mathcal{S}$ is not closed with respect to finite intersections. But the intersection of all topologies on M that contain $\mathcal{S}$ is a topology $\mathcal{T}$. It consists of all finite intersections of elements from $\mathcal{S}$ and their unions. It is the coarsest topology for which all f_i are continuous.

Definition 2.2.1

The topology $\mathcal{T}$ is the *topology induced* by ($f_i \mid i \in I$).

Theorem 2.2.2 (Universal Property of the Induced Topology)
The induced topology is the only topology on M with the property that any map $g: T \to M$ is continuous if and only if all compositions $f_i g: T \to X_i$ are continuous.

The proof of the last section can be adapted without major changes to this case.

Example 2.2.3
Let X and Y be topological spaces. The Cartesian product $M = X \times Y$ comes with two projection maps

$$X \xleftarrow{\mathrm{pr}_X} X \times Y \xrightarrow{\mathrm{pr}_Y} Y \; ,$$

which assign to a pair (x, y) its components x and y. The system $\mathcal{S}$ consists here of the strips of the form $\mathrm{pr}_X^{-1}(U) = U \times Y$ for open sets U of X and $\mathrm{pr}_Y^{-1}(V) = X \times V$ for open sets V of Y. Intersections of these strips yield *open rectangles*, and the induced topology consists of unions of such rectangles (see Fig. 2.4).
For metric spaces X and Y, these open sets are generated by the product metric

$$d((x, y), (x', y')) = \max\{d(x, x'), d(y, y')\}.$$

Fig. 2.4 The rectangles in the product topology

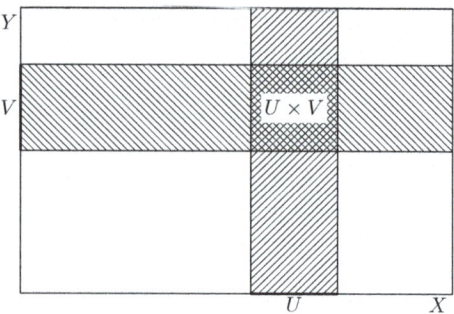

2.2 Products

Definition 2.2.4

The Cartesian product $\prod_{i \in I} X_i$ of sets X_i is the set of all I–tuples

$$(x_i \mid i \in I)$$

with $x_i \in X_i$. The topology induced on the Cartesian product by all projections

$$\operatorname{pr}_j : \prod_{i \in I} X_i \to X_j, \ (x_i \mid i \in I) \mapsto x_j$$

is called the *product topology*.

Example 2.2.5

An element in the product $\prod_{[0,1]} \mathbb{R}$ of copies of the real numbers over the index set $I = [0, 1]$ corresponds to an arbitrary function on the unit interval

$$f : [0, 1] \longrightarrow \mathbb{R}, \ x \longmapsto f(x).$$

The projection pr_x then evaluates f at x. A sequence (f_n) converges in the product space $\prod_{[0,1]} \mathbb{R}$ to an f (in the sense of Sect. 1.2) if and only if it converges pointwise, for each x. This is best seen using the universal property of the product. Let T be the subspace of $\mathbb{R}$ consisting of the numbers $1/n$ and their limit 0. Then a sequence (f_n) defines a map

$$g : T \longrightarrow \prod_{[0,1]} \mathbb{R}, \ 1/n \longmapsto f_n, \ 0 \longmapsto f.$$

If the sequence of functions converges pointwise, then the composition $\operatorname{pr}_x g$ in the diagram

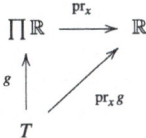

is continuous for all x because then the image of the sequence $1/n$ (and all other sequences in T that converge to zero) converges to $f(x)$. With the universal property, then g is also continuous, and thus the image (f_n) of the sequence $(1/n)$ under g is convergent. The converse is clear because images of convergent sequences under continuous maps, especially under the projection maps, are always convergent. Thus, the product topology on the set of all real maps generates the topology of *pointwise convergence*.

In the following definition, the concepts of 'subspace' and 'product' are combined. Since both are induced constructions, we can again expect a universal property.

Definition 2.2.6

If $p\colon X \to B$ and $q\colon Y \to B$ are two continuous maps into the same topological space B, then the subspace

$$X \times_B Y = \{(x, y) \in X \times Y \mid p(x) = q(y)\}$$

of the product space is called the *pullback* (or *fibre product*) of p and q.

The two projections of the product provide a diagram

$$\begin{array}{ccc} X \times_B Y & \xrightarrow{\mathrm{pr}_Y} & Y \\ {\scriptstyle \mathrm{pr}_X}\downarrow & & \downarrow{\scriptstyle q} \\ X & \xrightarrow{p} & B, \end{array}$$

which commutes, i.e., we have $p\,\mathrm{pr}_X = q\,\mathrm{pr}_Y$.

Examples 2.2.7

If B is a singleton space, then the pullback is the ordinary product. In this sense, the pullback generalises the ordinary product of two spaces. If $X = \{b\}$ is a point of B and p is the inclusion, then $\{b\} \times_B Y$ is homeomorphic to the subspace $q^{-1}(b)$ of Y, the *fibre* of q over b.

Pullbacks occur in many different situations. It is worth gaining familiarity with them, especially in view of later constructions. Here is their universal property.

Theorem 2.2.8 (Universal Property of the Pullback)

Let $f\colon T \to X$ and $g\colon T \to Y$ be continuous maps with $pf = qg$. Then there is a unique continuous map $T \to X \times_B Y$ that makes the diagram

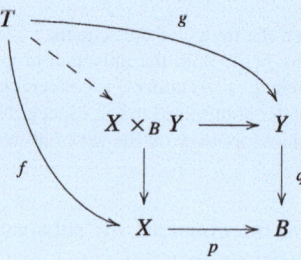

commute.

2.3 Sums

Proof. The map must be given by $t \mapsto (f(t), g(t))$. Continuity results directly from the universal properties of the product $X \times Y$ and the subspace $X \times_B Y$ of it. □

If $p: X \to B$ is a continuous map, then B is also called the *base* of p. If $f: B' \to B$ is continuous, then the map $B' \times_B X \to B'$ is called the *base change* of p along f. This process should not be confused with changing bases in linear algebra.

Exercises

Exercise 30 Associativity
The spaces $(X_1 \times X_2) \times X_3$ and $X_1 \times (X_2 \times X_3)$ are homeomorphic.

Exercise 31 Product of Open Sets
Is the product $\prod_{[0,1]}(0, 1)$ open in $\prod_{[0,1]} \mathbb{R}$?

Exercise 32 Product and Closure
Let $(X_i \mid i \in I)$ be a family of topological spaces and let $M_i \subseteq X_i$ for $i \in I$ be any subsets. Show the equality

$$\prod_{i \in I} \overline{M_i} = \overline{\prod_{i \in I} M_i}.$$

Exercise 33 Projections
Let X and Y be topological spaces. Is the projection

$$\mathrm{pr}_Y : X \times Y \longrightarrow Y$$

always closed? Is it always open?

Exercise 34 Product Rule
Let F be a subspace of X and G a subspace of Y. Do we have

$$\partial(F \times G) = (\partial F) \times G \cup F \times (\partial G)?$$

Exercise 35 Prime Spaces
If $\mathbb{R}$ is homeomorphic to a product $X \times Y$, then X or Y is a singleton.

2.3 Sums

The concept of the co-induced topology is dual to the idea of the induced topology. The term 'dual' here refers to the fact that we get from one concept to the other by reversing the arrow directions.

Let M be a set, let $(X_i \mid i \in I)$ be a family of topological spaces and

$$f_i : X_i \longrightarrow M$$

for each $i \in I$ a map. We seek a topology $\mathcal{T}$ on M for which all f_i are continuous. For this, the set

$$\mathcal{T}^{\mathrm{op}} = \bigcap_{i \in I} \{U \subseteq M \mid f_i^{-1} U \text{ is open in } X_i\}$$

of subsets of M must contain the topology $\mathcal{T}$. The system $\mathcal{T}^{\mathrm{op}}$ is itself a topology (check!) and thus the finest of its kind. A subset of M with respect to this topology is therefore open if and only if all of its pre-images are open.

Definition 2.3.1

The topology $\mathcal{T}^{\mathrm{op}}$ is called *co-induced by the family* $(f_i \mid i \in I)$.

Theorem 2.3.2 (Universal Property of the Co-Induced Topology)

The co-induced topology is the only topology on M with the property that every map $g \colon M \to T$ is continuous if and only if all compositions $g f_i \colon X_i \to T$ are continuous.

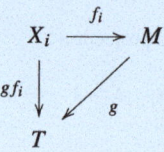

Proof. We have already mentioned that the maps f_i are continuous. With g, the compositions $g f_i$ are also continuous. Conversely, if the maps $g f_i$ are continuous, then for each open U in T the set $g^{-1}(U)$ is an element of $\mathcal{T}^{\mathrm{op}}$, because we have

$$f_i^{-1}(g^{-1}(U)) = (g f_i)^{-1}(U),$$

and thus g is continuous. To show uniqueness, let us consider for g the identity on M with respect to potentially different topologies. This works as already discussed for the induced topology. □

2.3 Sums

Fig. 2.5 The sum of three circles

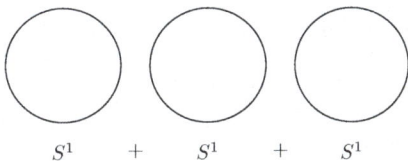

$S^1 \quad + \quad S^1 \quad + \quad S^1$

Definition 2.3.3

Let $(X_i \mid i \in I)$ be a family of topological spaces. Their *disjoint union* or *sum* is the set

$$\coprod_{i \in I} X_i = \{(x, i) \mid x \in X_i\}$$

together with the topology co-induced by the maps

$$\mathrm{in}_j \colon X_j \longrightarrow \coprod_{i \in I} X_i, \quad x \longmapsto (x, j).$$

Note that the images of these maps are pairwise disjoint, and their union is the whole set. For two summands X and Y, the notation $X + Y$ is often used for the sum.

A subset of $\coprod_i X_i$ is therefore open if it has the form $\coprod_{i \in I} U_i$ for a family of open subsets $U_i \subseteq X_i$. The maps in_j are open embeddings (see Fig. 2.5).

Example 2.3.4
The subspace $[0, 1] \cup {]2, 3[}$ of the real line $\mathbb{R}$ is homeomorphic to the sum $[0, 1] + {]2, 3[}$. Every open subset in the union is the disjoint union of an open subset of $[0, 1]$ with an open subset of $]2, 3[$.

▶ **Remark 2.3.5 (Universal Property of the Sum)** A map $g \colon \coprod X_i \to T$ is continuous if and only if all restrictions $g \, \mathrm{in}_j \colon X_j \to T$ are continuous.

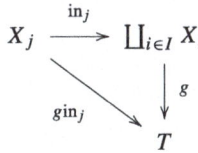

Exercises

Exercise 36 Discretion
If every space X_i is a singleton, then the sum $\coprod_{i \in I} X_i$ is homeomorphic to the set I with the discrete topology.

Exercise 37 Typical
Let $S^1 = \{z \in \mathbb{C} \mid \|z\| = 1\}$ and $q\colon S^1 \to S^1$ the map given by $q(z) = z^2$. Then the pullback $S^1 \times_{S^1} S^1$ of q along q is homeomorphic to $S^1 + S^1$.

Exercise 38 Sum of Open Sets
Is the sum $\coprod_{[0,1]}]0, 1[$ open in $\coprod_{[0,1]} \mathbb{R}$?

2.4 Identifications and Quotients

This section deals with the construction dual to embeddings.

Definition 2.4.1

Let X be a topological space. A surjection $p\colon X \to Y$ is called an *identification* if Y carries the topology co-induced by p.

Example 2.4.2

Let M be the set of all lines in $\mathbb{R}^{n+1}$ that pass through the origin 0. The space $\mathbb{R}^{n+1} \setminus 0$ maps surjectively onto M by assigning to a vector $x \neq 0$ the line spanned by x. The set M together with the co-induced topology is called *real projective space* and is denoted by $\mathbb{R}P^n$. Similarly, the set $\mathbb{C}P^n$ of lines in $\mathbb{C}^{n+1}$ becomes a topological space, a *complex projective space*.

Many examples of identifications arise from identifying (or 'glueing') points in topological spaces. To specify this statement, let us first recall the following terms.

Definition 2.4.3

A *partition* of M is a set $\mathcal{M}$ of subsets of M, such that

(Z1) $A \neq \emptyset$ for all $A \in \mathcal{M}$
(Z2) $\bigcup_{A \in \mathcal{M}} A = M$
(Z3) $A \neq B \implies A \cap B = \emptyset$ for all $A, B \in \mathcal{M}$

Definition 2.4.4

An *equivalence relation* on a set M is a subset $R \subseteq M \times M$ with the following properties:

(R1) reflexive: $R \supseteq \Delta_M = \{(x, x) \mid x \in M\}$
(R2) symmetric: $R \supseteq R^{-1} = \{(x, y) \mid (y, x) \in R\}$

2.4 Identifications and Quotients

(R3) transitive: $R \supseteq RR = \{(x, z) \mid \text{There is a } y \in M \text{ with } (x, y), (y, z) \in R.\}$

A partition $\mathcal{M}$ defines an equivalence relation

$$\bigcup_{A \in \mathcal{M}} A \times A,$$

and conversely, an equivalence relation R defines a partition

$$M/R = \{[x] \mid x \in M\}$$

into the *equivalence classes*

$$[x] = \{y \in M \mid (x, y) \in R\}.$$

In this way, partitions and equivalence relations correspond uniquely to each other.

Definition 2.4.5

The set M/R is called the *quotient* of M by the equivalence relation R. The map

$$\text{pr}\colon M \longrightarrow M/R, \ x \mapsto [x]$$

is called the *canonical projection*.

Definition 2.4.6

A *quotient space* of a topological space X is a quotient X/R of X together with the topology co-induced by the canonical projection.

Example 2.4.7
Define an equivalence relation on the interval $[0, 1]$ in which two elements are equivalent if they are equal or belong to the boundary $\{0, 1\}$. In this situation, we say that the equivalence relation is generated by the relation $0 \sim 1$. The quotient space is thus formed from the interval by identifying 0 with 1. This space is homeomorphic to the circle line because

$$[0, 1] \longrightarrow S^1, \ x \mapsto \exp(2\pi i x)$$

is a continuous map that factors over the quotient space because

$$\exp(2\pi i \cdot 0) = 1 = \exp(2\pi i \cdot 1),$$

and the induced map $[0, 1]/\sim \to S^1$ is an open continuous bijection.

Example 2.4.8
As another example, let X be the unit square $[0, 1] \times [0, 1]$ in $\mathbb{R}^2$. The vertical edges should be identified with each other in the reverse direction. The corresponding equivalence relation is thus

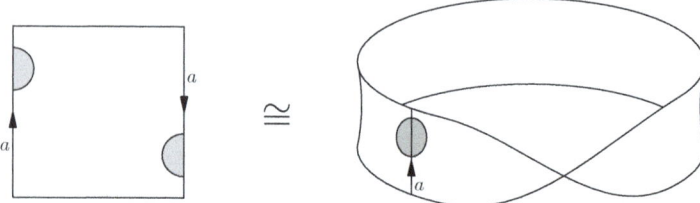

Fig. 2.6 The Möbius strip

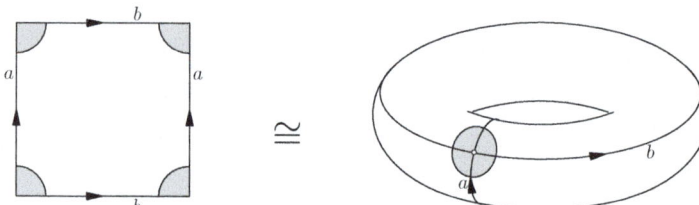

Fig. 2.7 The torus

generated by

$$(0, t) \sim (1, 1 - t)$$

for all t in the interval. The quotient space is called *Möbius strip* (see Fig. 2.6).

Example 2.4.9
As a final example, let X again be the unit square where opposite edges are identified. The equivalence relation is thus generated by

$$(0, t) \sim (1, t) \text{ and } (s, 0) \sim (s, 1)$$

for all s, t in the interval. The quotient space is the *torus* (see Fig. 2.7).

After this brief excursion about quotient maps, we return now to the general theory of identifications $p\colon X \to Y$. The following remark again provides a condition that determines the mapping behaviour of Y through the mapping behaviour of X.

▶ **Remark 2.4.10 (Universal Property of Identifications)** Let $p\colon X \to Y$ be an identification. A map $f\colon Y \to T$ is continuous if and only if the composition $fp\colon X \to T$ with p is continuous.

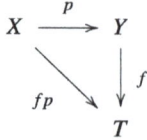

2.4 Identifications and Quotients

We often encounter continuous surjections, and we would like to recognise them as identifications. The following theorem provides a criterion for this.

Theorem 2.4.11
A continuous surjection that is open or closed is an identification.

Proof. If $f: X \to Y$ is continuous, then at least the pre-images of open sets are open. Let now V be a subset of Y with open pre-image. Then $f(f^{-1}(V))$ is an open subset of Y if f is open. Since f is surjective, we have $f(f^{-1}(V)) = V$. So, we see that V is open and f is an identification. If f is closed, we argue similarly. □

Example 2.4.12
Consider the surjection from the sphere to the projective space

$$p: S^n \longrightarrow \mathbb{R}P^n$$

that assigns to a vector of length one the line it generates. This map is open because if U is an open subset of $\mathbb{R}^{n+1} \setminus 0$ that contains no opposite pairs of points $\{x, -x\}$, then $p(U \cap S^n)$ is open because the pre-image of this set in the space $\mathbb{R}^{n+1} \setminus 0$ is open.

Example 2.4.13
The converse of the theorem is false. For example, we can divide the real line $\mathbb{R}$ into two equivalence classes, consisting of the two sets $A = \{x \in \mathbb{R} \mid x \leqslant 0\}$, and $U = \{x \in \mathbb{R} \mid x > 0\}$. The open sets of the quotient topology on $Y = \{A, U\}$ are $\emptyset$, $\{U\}$, and Y. This therefore provides an identification $\mathbb{R} \to \{A, U\}$ that is neither closed nor open: The image of $]-2, -1[$ is not open and the image of $[1, 2]$ is not closed.

Finally, we can introduce the concept dual to the pullback and some of its specialisations.

Definition 2.4.14

If $f: A \to X$ and $g: A \to Y$ are two maps with the same source, their *pushout* is constructed as the quotient space of $X + Y$ for the equivalence relation generated by

$$f(a) \sim g(a)$$

for $a \in A$.

The following characterisation of the pushout results directly from the universal properties of the sum and the quotient space.

Theorem 2.4.15 (Universal Property of the Pushout)
Let $p\colon X \to T$ and $q\colon Y \to T$ be continuous maps with $pf = qg$. Then there is a unique continuous map $X +_A Y \to T$ that makes the diagram

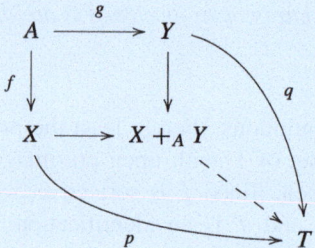

commute.

Examples 2.4.16
If $A = \emptyset$, the pushout coincides with the sum. If A is a subspace of a topological space X, we often write X/A for the pushout of the inclusion $A \subseteq X$ and the (unique) map from A to the one-point space:

$$\begin{array}{ccc} A & \xrightarrow{\subseteq} & X \\ \downarrow & & \downarrow \\ \star & \longrightarrow & X/A. \end{array}$$

We say the space X/A arises from X by *collapsing A to a point*. This is slightly irritating if A itself has no point; we have $X/\emptyset \cong X + \star$. Otherwise, the space X/A can also be seen as a quotient space of X according to the equivalence relation that identifies any two points from A with each other. A continuous map out of X/A is the same as a continuous map out of X that is constant on A. Examples of spaces obtained in this way are

$$[0, 1]/\{0, 1\} \cong S^1,$$

which was already considered above and more generally

$$D^n/S^{n-1} \cong S^n,$$

where

$$D^n = \{x \in \mathbb{R}^n \mid \|x\| \leqslant 1\}$$

is the n–dimensional *disk* (see Fig. 2.8).

2.4 Identifications and Quotients

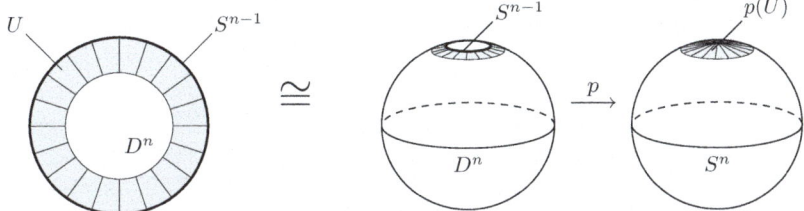

Fig. 2.8 The sphere S^n as a quotient D^n/S^{n-1}

A simple argument for why the continuous bijection indicated in the diagram is a homeomorphism is given in Sect. 4.1. So, there is a pushout diagram

$$\begin{array}{ccc} S^{n-1} & \overset{\subseteq}{\longrightarrow} & D^n \\ \downarrow & & \downarrow \\ \star & \longrightarrow & S^n. \end{array}$$

Similarly, we can show that there is also a pushout diagram

$$\begin{array}{ccc} S^{n-1} & \overset{\subseteq}{\longrightarrow} & D^n \\ {\scriptstyle\subseteq}\downarrow & & \downarrow \\ D^n & \longrightarrow & S^n. \end{array}$$

Here, the n-sphere is built from two parts, each homomorphic to D^n, that meet in a common equatorial S^{n-1}.

Definition 2.4.17

A continuous map $f \colon S^{n-1} \to X$ defines a space $X +_f D^n$ as the pushout of f and the inclusion map of the sphere S^{n-1} into the disk D^n. In particular, the diagram

$$\begin{array}{ccc} S^{n-1} & \overset{\subseteq}{\longrightarrow} & D^n \\ {\scriptstyle f}\downarrow & & \downarrow \\ X & \longrightarrow & X +_f D^n \end{array}$$

is a pushout diagram. We say that the space $X +_f D^n$ arises from X by *attaching an n-cell via* $f \colon S^{n-1} \to X$ (see Fig. 2.9).

It is immediately apparent that it is essential to understand maps of spheres in topological spaces to construct spaces by attaching cells. Conversely, we can study maps on spheres by considering the associated pushouts. By the way, the

Fig. 2.9 A new space arises from X by attaching an n–cell

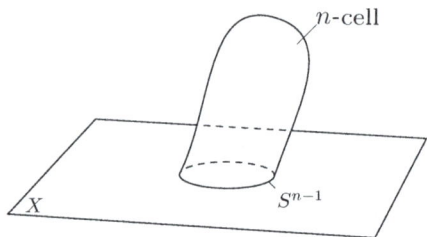

universal property of pushouts describes what a map $X +_f D^n \to Y$ is: a continuous map $g: X \to Y$ and a continuous extension of $gf: S^{n-1} \to Y$ to D^n. So, we immediately encounter the question of which maps on spheres can be extended to disks. We will discuss this in a later chapter. The attaching of cells can be iterated. Spaces that can be obtained by iteratively attaching cells have excellent properties. It follows directly from the definition that we know what it takes to define continuous maps out of them. Often, we try to 'resolve' arbitrary spaces through such nice spaces. When we frequently face the problem of constructing topological spaces with specific properties, we will usually also try to build them from cells.

Supplements

Quivers We can attach several cells at the same time. The simplest case is the following: A *quiver* $Q = (Q_0, Q_1, s, t)$ consists of a set Q_0 of *vertices*, a set Q_1 of *edges* as well as two maps $s, t: Q_1 \to Q_0$ that assign two vertices to each edge. A quiver is the same as a directed graph with loops and multiple edges. Each quiver has a *realisation* $|Q|$, which is defined as a pushout

$$\begin{array}{ccc} \coprod_{Q_1} \{0,1\} & \xrightarrow{\subseteq} & \coprod_{Q_1} [0,1] \\ {\scriptstyle (s,t)} \downarrow & & \downarrow \\ Q_0 & \longrightarrow & |Q|. \end{array}$$

The topological space $|Q|$ is a topological model for what the quiver Q describes as discrete data.

Surgery A construction closely related to the attaching of cells is surgery. It has the advantage that it leaves the 'dimension' of the space unchanged. It is based on the observation that the boundary of the 'square' $D^m \times D^n$ as a subset of $\mathbb{R}^m \times \mathbb{R}^n$ is the union

$$S^{m-1} \times D^n \cup D^m \times S^{n-1}.$$

2.4 Identifications and Quotients

Fig. 2.10 Surgery turns a 2–sphere into a torus

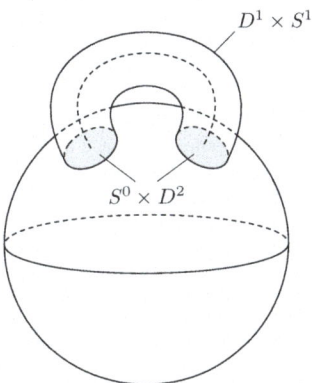

The two parts of the above decomposition should be thought of as thickened spheres: while these spheres can have different dimensions, the thickenings are of the same dimension. The intersection of the two parts is $S^{m-1} \times S^{n-1}$. This situation can now be exploited as follows. If X is a space, and

$$e \colon S^{m-1} \times D^n \longrightarrow X$$

is an embedding of such a thickened sphere, then we can remove the 'interior' of the image. A boundary is created that can be identified with $S^{m-1} \times S^{n-1}$ through e. Consequently, we can attach $D^m \times S^{n-1}$ to this boundary. The new space is thus created by *surgery on X using e*. It is worth considering this for $X = S^1 + S^1$ and embeddings $e \colon S^k \times D^{1-k} \to X$ with $k = 0, 1$ as well as for $X = S^2 + S^2$ and embeddings $e \colon S^k \times D^{2-k} \to X$ with $k = 0, 1, 2$ (see Fig. 2.10).

Surgery plays a significant role in differential topology. For example, it can be shown that a compact manifold without boundary is the boundary of a compact manifold if it is created by surgery from a sphere (see [Mil65]). There are also generalisations of this construction that, among other things, provide—in principle—a description of all 3– and 4–dimensional manifolds (see [Kir78] or [Kir89]).

Exercises

Exercise 39 Tinkering
Let $a < b < c < d$ be real numbers. Let $X = [a, d]$ and $A = [b, c]$. Show that X/A is also homeomorphic to a closed interval.

Exercise 40 Cones
Let $X \subseteq \mathbb{R}^n$ and $f \colon X \times [0, 1] \to \mathbb{R}^{n+1}$ be defined by

$$(x, t) \mapsto ((1 - t)x, t).$$

Describe the image Y of the map f. Then investigate whether f induces an identification $X \times [0, 1] \to Y$ when considering Y as a subspace of $\mathbb{R}^{n+1}$.

Exercise 41 Mapping Tori
If X is a topological space and $h \colon X \to X$ is a homeomorphism, then its *mapping torus* $T(X, h)$ is the quotient of the cylinder $[0, 1] \times X$ according to the equivalence relation given by $(1, x) \sim (0, h(x))$. Is there a pair (X, h) such that $T(X, h)$ is homeomorphic to the Möbius strip?

Exercise 42 Fixed Points
Let X be a topological space, let $h \colon X \to X$ be a homeomorphism and x a fixed point of h in X, i.e., we have $h(x) = x$. Show that there are then continuous maps between the circle S^1 and $T(X, h)$ such that the composition

$$S^1 \longrightarrow T(X, h) \longrightarrow S^1$$

is the identity.

Exercise 43 Multiple Twisted Möbius Strips
A Möbius strip is created by glueing the ends of a strip of paper, after half-twisting its ends. More generally, the $n/2$–*times twisted Möbius strip* is created by glueing a strip of paper, after twisting its ends $n/2$–times ($n \in \mathbb{Z}$). Into how many parts does the $n/2$–times twisted Möbius strip break when it is cut in the middle along the direction of the strip? Are the parts again twisted Möbius strips?

Exercise 44 Covers
Let $(U_j \mid j \in J)$ be an open cover of a topological space X. Then the map

$$\coprod_{j \in J} U_j \longrightarrow X$$

given by the inclusions is an identification. Is it open? Is it closed?

Exercise 45 Twisted Spheres
Let $h \colon S^{n-1} \to S^{n-1}$ be a homeomorphism. Then the pushout

$$\begin{array}{ccc} S^{n-1} & \stackrel{\subseteq}{\longrightarrow} & D^n \\ {\scriptstyle h}\downarrow & & \downarrow \\ S^{n-1} & & \\ {\scriptstyle \subseteq}\downarrow & & \downarrow \\ D^n & \longrightarrow & \Sigma^n(h) \end{array}$$

provides a space $\Sigma^n(h)$ homomorphic to S^n. There are diffeomorphisms h, so that $\Sigma^n(h)$ is homomorphic—but not diffeomorphic—to S^n (see [Mil56]).

2.4 Identifications and Quotients

Exercise 46 Equivalence Relations and Categories
Verify the statement: an equivalence relation is a small category with at most one morphism between any two objects, which must always be an isomorphism.

Exercise 47 Attaching
Show that the canonical map

$$X \longrightarrow X +_f D^n$$

is a closed embedding for every attaching map $f: S^{n-1} \to X$.

Exercise 48 Saturated or Not
Let $p: X \to Y$ be an identification. Show that the map $p: A \to p(A)$ induced by p that maps a subspace A of X to the subspace $p(A)$ of Y does not need to be an identification even if A is *saturated*, i.e., even if $A = p^{-1}p(A)$.

Connectivity and Separation

3

In this chapter, we examine the properties of topological spaces that can be transferred from one space to any other homeomorphic space. Such properties are called topological. The most important ones are connectivity, separation, and compactness. The first two are discussed in this chapter, and the last one is discussed in the next chapter.

3.1 Connectivity

Definition 3.1.1

A non-empty topological space X is called *connected* if there is no homeomorphism from X to a sum $X_1 + X_2$ of two non-empty topological spaces X_1 and X_2.

This definition explains the meaning of the term, but it does not help much when it comes to recognising connected spaces. The following theorem accomplishes this.

Theorem 3.1.2
For a topological space X, the following are equivalent.

- (Z) *The space X is connected.*
- (OC) *The two subsets $\emptyset$ and X of X are the only ones that are both open and closed.*
- (D) *Every continuous map from X to a discrete space is constant.*
- (S) *Every continuous map from X to $S^0 = \{-1, +1\}$ is constant.*

Proof. If A is an open and closed (we sometimes use the word 'clopen') subset of X, then X is the disjoint union of the open sets A and $X \setminus A$. In this case, a set in X is open if and only if its intersections with A and $X \setminus A$ are. It follows that X is homeomorphic to the sum of the subspaces A and $X \setminus A$. If X is assumed to be connected, then one of these spaces must be empty, and thus (OC) holds. This shows the first statement of the cycle. Next, assume that (OC) holds. Let f be a continuous map from X to a discrete space and let $x \in X$. Then $f^{-1}f(x)$ is a clopen and non-empty subset. Since this must coincide with all of X, the map f is constant. The conclusion from (D) to (S) is trivial. Finally, assume (S), and that X is homeomorphic to the sum of the spaces X_1 and X_2. Consider the continuous map

$$f : X \cong X_1 + X_2 \longrightarrow S^0,$$

that maps $x \in X_1$ to 1 and $x \in X_2$ to -1. Because f is constant, one of the two spaces X_1 or X_2 must be empty. □

It is immediately apparent that connectivity is a topological property: if a topological space is connected, then every homeomorphic one is so as well.

Example 3.1.3
Recall that a subset I of $\mathbb{R}$ is an *interval* if for any three real numbers with $a < b < c$ the following holds: if a and c are in I then so is b. Every interval I is connected because if

$$f : I \longrightarrow S^0$$

is a continuous map, and if $a < b$ are in I, Then according to the intermediate value theorem of calculus, we have $f(a) = f(b)$. So, the function f is constant. Conversely, if X is not an interval, then we can give a continuous map to S^0 that is not constant without much effort. So, the connected subspaces of $\mathbb{R}$ are the intervals. Since intervals are connected, connectivity properties of topological spaces are often examined using intervals. This approach leads to the concept of path connectivity, which is discussed in detail in Sect. 6.2.

Example 3.1.4
We consider the subspace

$$X = (\mathbb{Q} \times [0, +\infty[) \cup ((\mathbb{R} \setminus \mathbb{Q}) \times]-\infty, 0[)$$

of $\mathbb{R}^2$ and want to show that it is connected. Suppose we have a decomposition of the space into closed subsets X_1 and X_2, and let $(r, 0) \in X_1$. Then the entire interval $\{r\} \times [0, \infty[$ is in X_1 because it is connected and therefore cannot be decomposed. Since X_1 is open there is a whole vertical strip in X_1 (Fig. 3.1).
Similarly, we can see that X_2 also must consist of vertical strips. If one of the two sets is not empty then they touch each other in the upper or lower area, which contradicts the openness of the sets.

Now, we will prove some statements allowing us to construct more examples of connected spaces from known ones.

3.1 Connectivity

Fig. 3.1 Vertical strips

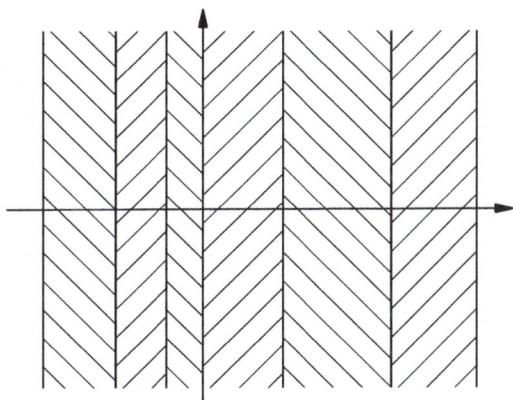

Corollary 3.1.5
Let $f: X \to Y$ be a continuous surjection. If X is connected, then so is Y.

Proof. If $g: Y \to S^0$ is continuous, so is the composite $fg: X \to S^0$. Therefore, this is constant. Since f is surjective, the map g must be constant. □

In particular, the image of an interval I under a continuous function $I \to \mathbb{R}$ is again an interval. Therefore, the intermediate value theorem of calculus appears here as a particular case.

Corollary 3.1.6
Given a family of connected subspaces of a topological space that are pairwise not disjoint, then its union is connected.

Proof. Let $(Z_i \mid i \in I)$ be the family of subspaces and

$$f: \bigcup_{i \in I} Z_i \longrightarrow S^0$$

a continuous map. Then its restriction to each subspace Z_i is constant. Since the pairwise intersections are not empty, the map f must be constant. □

Corollary 3.1.7
Let X be a topological space and $x \in X$. Among all connected subspaces of X that contain x, there is one largest, and this is closed.

Proof. According to the last conclusion, the union $Z(x)$ of all connected subspaces that contain x is connected and maximal among these. It is also closed because if

$$f: \overline{Z(x)} \longrightarrow S^0$$

is a continuous map, then its restriction to $Z(x)$ is constant. The pre-image of this value is a closed subset of $\overline{Z(x)}$. Since this is also closed in X and contains $Z(x)$ the closure of $Z(x)$ is also included. Therefore, the function f must be constant. □

Definition 3.1.8

The maximum connected set $Z(x)$ that contains x is called the *connected component* of x.

Example 3.1.9
Let $\mathbb{Q} \subset \mathbb{R}$ be the subspace of rational numbers and let $r \in \mathbb{Q}$. If $r' \in Z(r)$ is another rational number and q is an irrational number between r and r' then

$$f: Z(r) \longrightarrow S^0$$
$$x \mapsto \begin{cases} -1 & \text{for } x < q \\ +1 & \text{for } x > q \end{cases}$$

is a non-constant continuous map. Therefore, we have $Z(r) = \{r\}$. By the way, spaces whose connected components consist of only one point are called *totally disconnected*.

The example shows that spaces do not have to be the topological sum of their connected components because $\mathbb{Q}$ is not discrete. On the other hand:

Corollary 3.1.10
If there are only finitely many connected components in X, then X is their topological sum.

Proof. Each component of X is open because its complement is the finite union of the other connected components, and these are closed. □

Fig. 3.2 Points in the product space

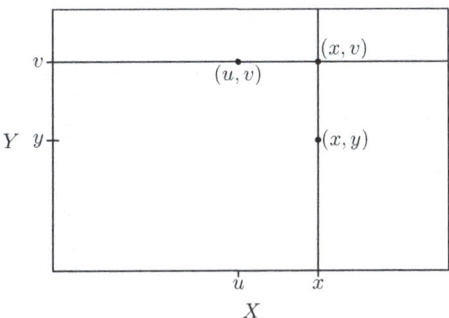

Corollary 3.1.11
A product $X \times Y$ is connected if and only if both factors X and Y are.

Proof. One direction is simple because X and Y are continuous images of $X \times Y$. Now let X and Y be connected. Then we choose any point (x, y) in the product and consider the connected component $Z(x, y)$. It is, therefore, sufficient to show that this is all of $X \times Y$. Let (u, v) be any point of the product (see Fig. 3.2). Then

$$(X \times \{v\}) \cup (\{x\} \times Y)$$

is a connected subspace of the product that contains (x, y) and (u, v). Both parts are homeomorphic to X or Y, so connected by assumption. Their intersection contains (x, v), so it is not empty. With this space, the point (u, v) is also in $Z(x, y)$. □

Supplement

Locally Connected A topological space X is called *locally connected* if every neighbourhood of a point still includes a connected one. For example, the space $\mathbb{R} \setminus 0$ is locally connected but not connected. The *sine space*

$$X = 0 \times \mathbb{R} \cup \{ (x, \sin(1/x)) \in \mathbb{R}^2 \mid x > 0 \} \subset \mathbb{R}^2$$

is connected, but not locally connected (Fig. 3.3).

Locally connected spaces have open connected components.

Fig. 3.3 The sine space

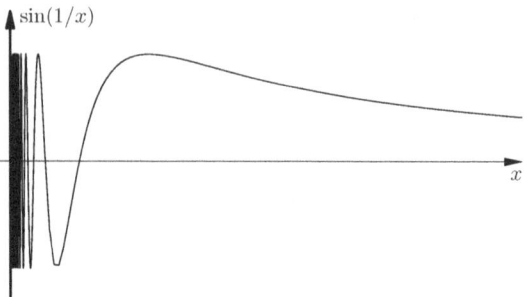

Exercises

Exercise 49 Cutting
Let A be a subset and Z a connected subset of X. Show: if Z intersects both A and $X \setminus A$ then Z intersects the boundary ∂A.

Exercise 50 Touching
Let M be a connected subspace of a topological space X and $\overline{M}$ the closure of M in X. If $M \subseteq N \subseteq \overline{M}$ then N is also connected.

Exercise 51 Not Without Counterexample
Let M and N be connected subspaces of a topological space X. Must $M \cap N$ then also be connected? And $M \cup N$?

Exercise 52 Connecting
Let X be a topological space. Suppose that for any two points in X there is a continuous map $\mathbb{R} \to X$ that hits both. Then X is connected.

Exercise 53 Gracious
Prove that for all $X \subseteq \mathbb{R}$ and all $f: X \to \mathbb{R}$ the following statements are equivalent.

  (I) The set X is an interval and f is continuous.
  (G) The graph of f is both connected and locally connected (see supplement).

3.2 Separation and Continuous Extendability

Separation properties ensure the existence of sufficiently many open sets to separate certain subsets of the space from each other. In this section, we show that they also ensure the existence of certain continuous real-valued functions that are used in various places in mathematics.

3.2 Separation and Continuous Extendability

Fig. 3.4 The separation axioms

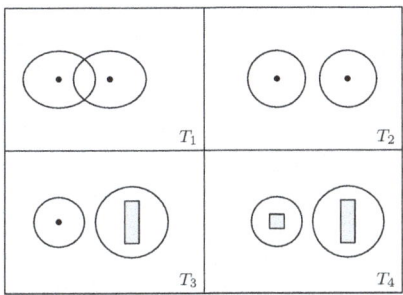

Definition 3.2.1

We define the following properties of a topological space X (see Fig. 3.4).

(T1) Any two different points of X have neighbourhoods not containing the other point.
(T2) Any two different points of X have disjoint neighbourhoods.
(T3) Any point and closed set not containing that point have disjoint neighbourhoods. (Earlier, we have only defined neighbourhoods of points. A neighbourhood of a subset is a superset of an open subset that contains the subset.)
(T4) Any two closed disjoint subsets of X have disjoint neighbourhoods.

A topological space X is a *Hausdorff space* if it has property (T2). If a topological space fulfils all separation properties, it is called *normal*.

The Hausdorff property will play a crucial role in the context of compactness in the next chapter.

Examples 3.2.2
Every metric space is a Hausdorff space. To see this, note that the ε–neighbourhoods of two different points are disjoint if ε is at most half the distance between the two points. Every Hausdorff space also fulfils (T1). The first separation property is equivalent to the statement that every singleton subset of X is closed (exercise!). If we equip the natural numbers with the 'complements-of-finite-subsets topology', then points are closed, but all other separation properties do not hold.

At this point, we refer to the book [SS95], in which we can find many examples of topological spaces that only have some separation properties. However, we do not want to get lost in such subtleties. If we demand a separation property it will usually be the Hausdorff property, sometimes also normality. We have already seen that metric spaces have the Hausdorff property. In fact, they are even normal. To see this, we introduce another term first.

Fig. 3.5 A Urysohn function

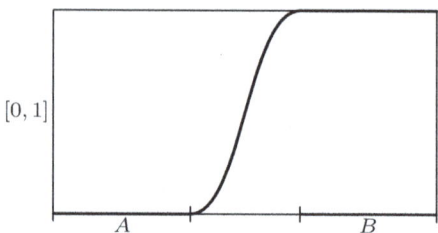

Definition 3.2.3

Let A and B be closed disjoint subsets of a topological space X. A continuous function $f\colon X \to [0,1]$ is called a *Urysohn function* for A and B if f is constantly 0 on A and constantly 1 on B (Fig. 3.5).

Urysohn functions are helpful when we want to decompose a real function g on X. For example, if $X = U \cup V$, with U, V open, and f a Urysohn function for $A = X \setminus U$ and $B = X \setminus V$, then

$$g = fg + (1-f)g$$

provides a decomposition of g into continuous functions, each of which vanishes outside of U or V. Of course, instead of two sets, we can also consider finitely many open subsets. This can be particularly useful when studying functions on manifolds.

Example 3.2.4
Let X be a metric space. Then there is a Urysohn function for any two disjoint closed subsets: if we define the distance of a point to the set A by

$$d_A(x) = \inf_{a \in A} d(x,a),$$

then $f = d_A/(d_A + d_B)$ has the desired property. Here, we note that the distance function d_A outside of A (or B) is strictly positive and the denominator of f therefore never vanishes.

If we have a Urysohn function f, we also get separating neighbourhoods of the closed sets A and B by

$$f^{-1}[0, 1/2[\quad \text{and} \quad f^{-1}]1/2, 1].$$

In particular, it follows that metric spaces have the property (T4). The following theorem states that (T4) is also sufficient for the existence of Urysohn functions. Thus, in (T4) spaces, the function defined on $A \cup B$ can be continuously extended to X. It also states that extensions of all other real-valued functions on closed subsets exist.

3.2 Separation and Continuous Extendability

> **Theorem 3.2.5 (Tietze–Urysohn)**
> *For every topological space X, the following are equivalent:*
>
> (T4) *The space X satisfies (T4).*
> (UE) *For any two disjoint closed subsets A and B of X, there exists a Urysohn function.*
> (TE) *For $A \subseteq X$ closed and $f: A \to \mathbb{R}$ continuous, there exists a continuous function $F: X \to \mathbb{R}$ that agrees with f on A.*

Proof. In (T4) spaces, for any two closed disjoint A and B, a set C can be found that fulfils

$$A \subseteq \overset{\circ}{C} \subseteq \bar{C} \subseteq X \setminus B.$$

To do this, we only need to separate A and B by disjoint neighbourhoods U and V, and then choose $U = C$. Since this works so well, we can insert a set as often as we like. A chain

$$\mathcal{A} = (A = A_0 \subseteq A_1 \subseteq \cdots \subseteq A_k \subseteq X \setminus B)$$

is called *admissible* if always $\bar{A}_i \subseteq \overset{\circ}{A}_{i+1}$ and $\bar{A}_k \subseteq X \setminus B$ applies. Each admissible chain can be refined in X to an admissible chain

$$A_0 \subseteq A'_0 \subseteq A_1 \subseteq A'_1 \subseteq \cdots \subseteq A_k \subseteq A'_k \subseteq X \setminus B$$

of double length. This way, we can find a sequence $(\mathcal{A}_n \mid n \in \mathbb{N})$ of admissible chains where $\mathcal{A}_{n+1}$ is a refinement of double length of $\mathcal{A}_n$ and $\mathcal{A}_0 = (A \subseteq X \setminus B)$. Given $\mathcal{A}_n$, we define a staircase function by

$$f_n(x) = k2^{-n} \text{ for } x \in A_k \setminus A_{k-1},$$

where $A_{-1} = \emptyset$ and $A_{2^n} = X$ is set. The sequence of functions is monotonically decreasing and bounded. It thus converges pointwise to a limit function f. On A, the map f is constantly 0 and, on B, constantly 1 because this is true for each f_n. The astonishing thing now is that f is continuous and thus a Urysohn function, although none of the f_n has to be. To see this, note

$$|f_n(x) - f_{n+1}(x)| \leqslant 2^{-(n+1)}$$

and, with respect to the chain $\mathcal{A}_n$,

$$|f_n(x) - f_n(y)| \leqslant 2^{-n}$$

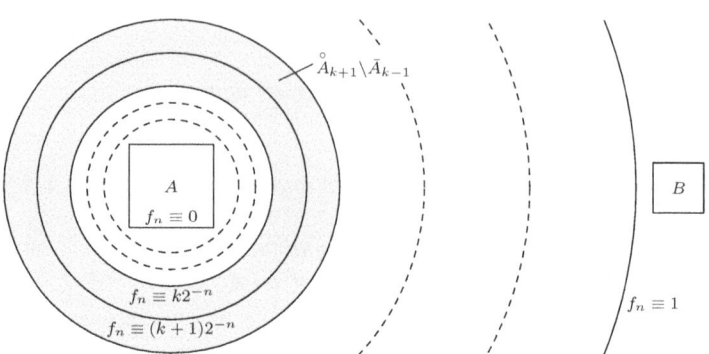

Fig. 3.6 On the continuity of f

holds for all $x, y \in \overset{\circ}{A}_{k+1} \setminus \overline{A}_{k-1}$. It follows

$$|f(x) - f(y)| \leqslant |f(x) - f_n(x)| + |f_n(x) - f_n(y)| + |f_n(y) - f(y)|$$

$$\leqslant \sum_{k=n+1}^{\infty} 2^{-k} + 2^{-n} + \sum_{k=n+1}^{\infty} 2^{-k} = 3 \cdot 2^{-n}.$$

If we choose an n that satisfies $\varepsilon > 3 \cdot 2^{-n}$ then x can be varied in the open neighbourhood $\overset{\circ}{A}_{k+1} \setminus \overline{A}_{k-1}$ without getting fluctuations larger than ε (see Fig. 3.6).

Now, we show that the Urysohn property (UP) implies the Tietze property (TP). First, let $f \colon A \to [-1, 1]$ be a continuous function that maps into the bounded interval. Choose a Urysohn function G on X for the disjoint closed subsets $f^{-1}[-1, -1/3]$ and $f^{-1}[1/3, 1]$. The function $F_1 = 2/3 G - 1/3 \colon X \to \mathbb{R}$ then satisfies

$$|f(x) - F_1(x)| \leqslant 2/3 \text{ for } x \in A$$

$$|F_1(x)| \leqslant 1/3 \text{ for } x \in X.$$

This is best checked with a case distinction of the form

$$x \in f^{-1}[-1, -1/3], \quad x \in f^{-1}]-1/3, +1/3[, \quad x \in f^{-1}[1/3, 1].$$

The function F_1 indeed is defined on X but only approximately agrees with f on A. To obtain a better approximation, the method is applied to the error function

$$f' \colon A \longrightarrow [-2/3, 2/3], \quad x \longmapsto f(x) - F_1(x)$$

3.2 Separation and Continuous Extendability

and a function $F_2: X \to \mathbb{R}$ is obtained that satisfies

$$|f'(x) - F_2(x)| \leqslant (2/3)^2 \text{ for } x \in A$$
$$|F_2(x)| \leqslant 1/3 \cdot 2/3 \text{ for } x \in X$$

Continuing in this way, a sequence of functions $(F_n \mid n \in \mathbb{N})$ is created with

$$\left|f(x) - \sum_{j=1}^{n} F_j(x)\right| \leqslant (2/3)^n \text{ for } x \in A$$
$$|F_n(x)| \leqslant 1/3 \cdot (2/3)^{n-1} \text{ for } x \in X.$$

The series $\sum_n F_n$ converges uniformly to a continuous function F that agrees with f on A. Thus, we have (TE) at least for bounded functions. For the general case of a function f, the homeomorphism

$$\arctan: \mathbb{R} \longrightarrow {]-\pi/2, +\pi/2[}$$

can be used to transform the function f into a bounded function: with what has already been proven, the function

$$(2/\pi) \cdot \arctan(f): A \longrightarrow [-1, +1]$$

can be extended to a function $F': X \to [-1, +1]$. Unfortunately, an inverse transformation may not be available because the values -1 and 1 can be in the image of F'. To circumvent this problem, choose a Urysohn function G that has the value 1 on A and vanishes on $F'^{-1}\{-1, 1\}$. The desired extension of f is then given by the formula

$$F = \tan(\pi/2 \cdot G \cdot F').$$

The proof is thus complete as it was already shown before how to get from (UE) and thus from (TE) to (T4). □

As a small application of the theorem, we will show a result about complements of closed subsets of Euclidean spaces. First, it should be noted that there are homeomorphic closed subsets of $\mathbb{R}^n$ that do not have homeomorphic complements. Examples are given by the trivial knot, the trefoil knot, and the figure-eight knot (see Sect. 2.1), which have pairwise non-homeomorphic complements (even if that might not be so easy to see at the moment). In fact, it is true that the complement determines a knot (see [GL89]). However, the following result shows that the complement of a closed subset only depends on it in a stable sense.

Theorem 3.2.6

Let $A \subseteq \mathbb{R}^a$ and $B \subseteq \mathbb{R}^b$ be two closed subsets that are homeomorphic. Then the complements of $A \cong A \times 0$ and $B \cong 0 \times B$ in $\mathbb{R}^{a+b} \cong \mathbb{R}^a \times \mathbb{R}^b$ are homeomorphic.

Proof. Let $\varphi \colon A \to B$ be a homeomorphism. According to the theorem of Tietze–Urysohn, there is an extension $\Phi \colon \mathbb{R}^a \to \mathbb{R}^b$. Correspondingly, let $\Psi \colon \mathbb{R}^b \to \mathbb{R}^a$ be an extension of the inverse ψ of φ. Then

$$L \colon \mathbb{R}^a \times \mathbb{R}^b \longrightarrow \mathbb{R}^a \times \mathbb{R}^b, \quad (x, y) \longmapsto (x, y - \Phi(x))$$

and

$$R \colon \mathbb{R}^a \times \mathbb{R}^b \longrightarrow \mathbb{R}^a \times \mathbb{R}^b, \quad (x, y) \longmapsto (x - \Psi(y), y)$$

are homeomorphisms with obvious inverses. Here, the homeomorphism L has the property that it maps the graph Γ of φ homeomorphically onto $A \times 0$. And R maps Γ homeomorphically onto $0 \times B$. Thus, a desired homeomorphism results from the composite RL^{-1} restricted to the complement of $A \times 0$ with values in the complement of $0 \times B$. □

Finally, it should be noted that not all continuous functions can be extended when the target area is changed. For example, the intermediate value theorem of analysis says that there is no extension of the identical map from $\{0, 1\}$ to $[0, 1]$:

$$\begin{array}{ccc} \{0,1\} & \xrightarrow{\subseteq} & [0,1] \\ {\scriptstyle \text{id}}\downarrow & \swarrow & \\ \{0,1\} & & \end{array}$$

In order to extend functions to $\mathbb{R}$ the closedness condition in (TE) cannot be waived: for example, the function $f \colon \mathbb{R} \setminus 0 \to \mathbb{R}$ that is constant 0 for negative numbers and constant 1 otherwise does not allow a continuous extension.

Supplement

Regularity A topological space X is called *regular* if it fulfils the properties (T1) and (T3). It is called *completely regular* (T3.5) if for every $x \in X$ and every closed subset A of X there exists a continuous $f \colon X \to [0, 1]$ that fulfils $f(x) = 0$ and is constant 1 on A. Completely regular spaces are regular but not necessarily

3.2 Separation and Continuous Extendability

normal. Tychonoff was able to show that these spaces are up to homeomorphism the subspaces of a cuboid $\prod_I [0, 1]$ for suitable index sets I, see for example [Eng68].

Exercises

Exercise 54 Inheritance
Show that the Hausdorff property is inherited by subspaces and products. Also, show this for regularity.

Exercise 55 Discretion
A finite topological space is a Hausdorff space if and only if it is discrete.

Exercise 56 Punctuality
A topological space X is a Hausdorff space if and only if for every point x the intersection of its closed neighbourhood equals $\{x\}$. In particular, the singleton subsets are then closed. Is there a topological space whose singleton subsets are closed but which is not a Hausdorff space?

Exercise 57 The Whole Zariski
Is $\mathrm{Spec}(\mathbb{Z})$ a Hausdorff space?

Exercise 58 Separated Maps
Let $f: X \to Y$ be continuous. The image of the diagonal

$$\Delta: X \longrightarrow X \times_Y X$$

is closed if and only if any two points $x \neq x'$ with $f(x) = f(x')$ have disjoint open neighbourhood in X. Such maps are called *separated*. The fibres of separated maps are Hausdorff spaces. (Consider the case $Y = \star$ first?)

Exercise 59 Generalised Points
Let $f: X \to B$ be separated. If $s: B \to X$ is a section of f, so that $fs = \mathrm{id}_B$, then s is a closed embedding. This generalises the fact that singleton subsets of Hausdorff spaces are closed.

Exercise 60 Separation Algebra
Assume X has the properties (T2) and (T3) and A is closed in X. Does X/A then fulfil property (T2)?

Exercise 61 Surjections
Let $f: X \to Y$ be a surjective, continuous map. Must Y be normal if X is normal?

Exercise 62 Separated Lines
Show that the projective spaces are Hausdorff spaces.

Exercise 63 The Better Hausdorff
Show that in regular spaces (see supplement) any two points can be separated by open neighbourhoods whose closures are still disjoint.

Compactness and Mapping Spaces

4

In this section, we introduce the concept of compactness of topological spaces. After that, we discuss a relative version of this notion, that of a proper map. A technical section about Tychonoff's theorem follows, which can be skipped at the first reading. We then turn to mapping spaces. As it is mainly the function spaces that are of interest in analysis, in topology, it is the spaces of continuous maps that provide extremely important examples of topological spaces. We conclude the chapter with a technical section about the category of compactly generated spaces, which can also be skipped initially.

4.1 Compactness

From the analysis course, the reader may remember that 'compact' has something to do with 'closed and bounded'. That is true, and the corresponding theorem can be found in Sect. 4.2. However, set-theoretical topology is based on the open sets. Therefore, we define the concept first in these terms.

> **Definition 4.1.1**
>
> Let X be a topological space. A family $(U_j \mid j \in J)$ of open subsets of X is an *open cover*, if every point of X is contained in one of the sets U_j. If I is a subset of the index set J, so that $(U_j \mid j \in I)$ is also an open cover, we speak of a *subcover*. An open cover is called *finite* or *countable* if this applies to the index set. If every open cover of X has a finite subcover, then X is called *compact*.
>
> Compactness is obviously a topological invariant, so if two spaces are homeomorphic, one is compact if and only if the other is.

Example 4.1.2
The interval $[0, 1]$ is compact, even though that is not obvious with the definition just given. Therefore, we should also justify it. Let $(U_j \mid j \in J)$ be an open cover of $[0, 1]$. We consider the subset

$$M = \{s \in [0, 1] \mid \text{There is a finite subcover of } [0, s].\}$$

of $[0, 1]$. The set M is not empty because the point 0 is in it. Furthermore, if s is in M, then the whole interval $[0, s]$ is contained in it. Now let t be the supremum of M. Then there is a set in the cover that contains t. This set then also contains a point s from M. This set and the finitely many sets that cover $[0, s]$ together cover $[0, t]$. Thus, the point t is in M. The argument also shows that t cannot be smaller than 1 because otherwise, there would be points in M larger than t. Therefore, we have $t = 1$. Because t is in M, there is then a finite subcover of $[0, 1]$. Thus, the space $[0, 1]$ is compact.

Before we discuss further examples, we should convince ourselves that compactness is a valuable property of topological spaces. The following theorems show this.

> **Theorem 4.1.3**
> *Every closed subset of every compact space is compact as a subspace.*

Proof. Let A be closed in the compact space X. If $(U_j \mid j \in J)$ is an open cover of A, then there are, by the construction of the subspace topology, open subsets V_j in X that satisfy $U_j = A \cap V_j$. The sets V_j do not necessarily form an open cover of X, but we obtain such a cover if we add the open subset $X \setminus A$ of X. By assumption, then only finitely many of these are needed to cover X. The corresponding finitely many U_j cover A. □

> **Theorem 4.1.4**
> *Every compact subspace of a Hausdorff space is closed in it.*

Proof. Let X be a Hausdorff space and K a compact subspace. It suffices to show that the complement $X \setminus K$ is open in X. If x is a point in this complement, then for each point k of K we can choose two disjoint open neighbourhoods U_k and V_k of k and x. Then $(U_k \mid k \in K)$ is an open cover of K. According to the assumption, there is, therefore, a finite subcover $(U_k \mid k \in K_0)$ of it. But then

$$\bigcap_{k \in K_0} V_k$$

is an open neighbourhood of x that does not intersect K. Because one such exists for every x from $X \setminus K$, the complement of K is open in X (see Fig. 4.1). □

4.1 Compactness

Fig. 4.1 For the proof of Theorem 4.1.4

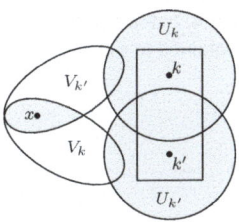

Theorem 4.1.5
The image of every compact space under any continuous map is a compact subspace.

Proof. Let X be a compact space and $f: X \to Y$ a continuous map into another space Y. Then it is claimed that $f(X)$ is a compact subspace of Y. Let therefore $(U_j \mid j \in J)$ be an open cover of $f(X)$. By construction of the subspace topology, there are open subsets V_j of Y with $U_j = f(X) \cap V_j$. The pre-images provide an open cover $(f^{-1}(V_j) \mid j \in J)$ of X. According to the assumption, it has a finite subcover $(f^{-1}(V_j) \mid j \in J_0)$. It is now sufficient to show that $(U_j \mid j \in J_0)$ is an open cover of $f(X)$. This can be seen as follows: If y is in $f(X)$, then there is a x in X with $y = f(x)$. This point x is in a $f^{-1}(V_j)$ with a j from J_0. But then y is in $f(f^{-1}(V_j)) = V_j \cap f(X) = U_j$ with this j from J_0. □

For the next consequence, each of the three preceding theorems is used.

Corollary 4.1.6
A continuous map from a compact space into a Hausdorff space is closed.

Proof. Let X be compact, let Y be a Hausdorff space, and $f: X \to Y$ continuous. If A is closed in X, then A is compact, so $f(A)$ is compact. Then $f(A)$ is closed. □

Corollary 4.1.7
A continuous bijection from a compact space into a Hausdorff space is a homeomorphism.

Proof. According to the previous conclusion, the map is closed. This means the inverse map is continuous because pre-images of closed sets are closed. □

Fig. 4.2 Each box is contained in a cover set

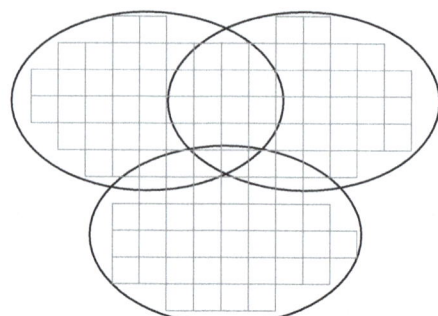

For the upcoming theorem, we assume that X is a metric space. For such spaces, compactness can also be expressed differently.

Definition 4.1.8

A $\delta > 0$ is called a *Lebesgue number* of an open cover $\mathcal{U}$ of X, if for every $x \in X$ there exists a $U \in \mathcal{U}$ that contains the δ–neighbourhood of x.

In the Fig. 4.2, each box is contained in one of the cover sets.
Lebesgue numbers will be helpful later in the construction of continuous maps. The significance of the following notion will only become apparent through the subsequent theorem.

Definition 4.1.9

A metric space X is called *totally bounded*, if for each $\varepsilon > 0$ there exists a finite set $E \subseteq X$, so that $\{U_\varepsilon(a) \mid a \in E\}$ is a cover.

Theorem 4.1.10
For metric spaces X, the following statements are equivalent.

(K) *The space X is compact.*
(PK) *Every continuous real-valued function on X is bounded.*
(TL) *The space X is totally bounded, and for every open cover of X there is a Lebesgue number.*

Proof. First, it is clear that every real-valued function f on a compact X must be bounded because $f(X)$ is compact. If $\mathcal{U}$ is an open cover of X, the assignment

$$f: X \longrightarrow \mathbb{R}, \quad x \longmapsto \sup_{U \in \mathcal{U}} \{\min\{1, d_{X \setminus U}(x)\}\}$$

4.1 Compactness

is a strictly positive real function. Here, we use d_A to denote the distance function to the set A (see Sect. 3.2). The map f is continuous, even *contracting*, as we have

$$|f(x) - f(y)| \leq d(x, y)$$

for all x, y in X. This can be seen by applying the infimum and minimum to the triangle inequality

$$d(x, z) \leq d(x, y) + d(y, z)$$

and rearranging the inequality. The property (PK) for $1/f$ thus provides us with a $\delta > 0$ with

$$f(x) \geq \delta$$

for all x. This number is a Lebesgue number, because if $x \in X$ and $U \in \mathcal{U}$ is chosen such that $d_{X \setminus U}(x) \geq \delta$, then $U_\delta(x) \subseteq U$. Total boundedness also follows from (PK). Otherwise, there would be an $\varepsilon > 0$ and a sequence (a_n) of elements from X with $d(a_n, a_m) \geq \varepsilon$. Let f be the function

$$f(x) = \begin{cases} \frac{3n}{\varepsilon}(\frac{\varepsilon}{3} - d(x, a_n)) & \text{for } x \in U_{\frac{\varepsilon}{3}}(a_n) \\ 0 & \text{otherwise.} \end{cases}$$

Then f is continuous but not bounded because $f(a_n) = n$. Finally, compactness follows directly from (TL) by taking a Lebesgue number δ for a cover $(U_j \mid j \in J)$ and then using the total boundedness to choose a finite subset E for δ, so that $(U_j \mid j \in E)$ becomes a finite subcover. □

A topological space, on which every continuous function is bounded, is also referred to as *pseudocompact*. For metric spaces, this is equivalent to their compactness.

Finally, we introduce a local version of the concept of compactness. It will be particularly useful in our treatment of mapping spaces.

Definition 4.1.11

A topological space is called *locally compact* if every neighbourhood of every point contains a compact neighbourhood of the point.

Examples 4.1.12
The space $\mathbb{R}^n$ is locally compact, but also every open subset of $\mathbb{R}^n$. A more general class of examples is provided by the following theorem.

Fig. 4.3 For the proof of Theorem 4.1.13

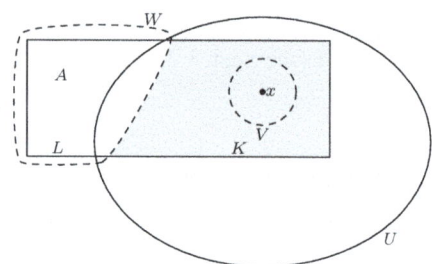

Theorem 4.1.13
Let X be a Hausdorff space whose points have compact neighbourhoods. Then X is locally compact.

Proof. Let x be a point of X and U an open neighbourhood of it. We must find a compact neighbourhood $K \subseteq U$ of X (see Fig. 4.3). By assumption, there is, in any case, a compact neighbourhood L. Then $L \cap U$ is open in L. The complement A is closed in L, thus compact. There are then open, disjoint subsets V and W of X with $x \in V$ and $A \subseteq W$. Then $K = L \setminus W$ is closed in L, thus compact. Furthermore K is in $L \setminus A = L \cap U$, thus in U. And because $L \cap V \subseteq L \setminus W = K$, it also contains a neighbourhood of x. □

Corollary 4.1.14
All compact Hausdorff spaces are locally compact.

Figure 4.4 provides an overview of the implications between different concepts of compactness. Here, the space $\mathbb{Q}_+$ is the one-point compactification of $\mathbb{Q}$, a construction that will be explained in the following supplements. The space $\{0, 1\}_{cl}$ is the two-point set $\{0, 1\}$ with the clump topology.

Figure 4.5 provides an overview of implications between various separation properties (see also the following exercise 'Everything normal').

Supplements

Sequential Compactness A topological space X is called *sequentially compact*, if every sequence in X has a convergent subsequence. Compact spaces need not be

4.1 Compactness

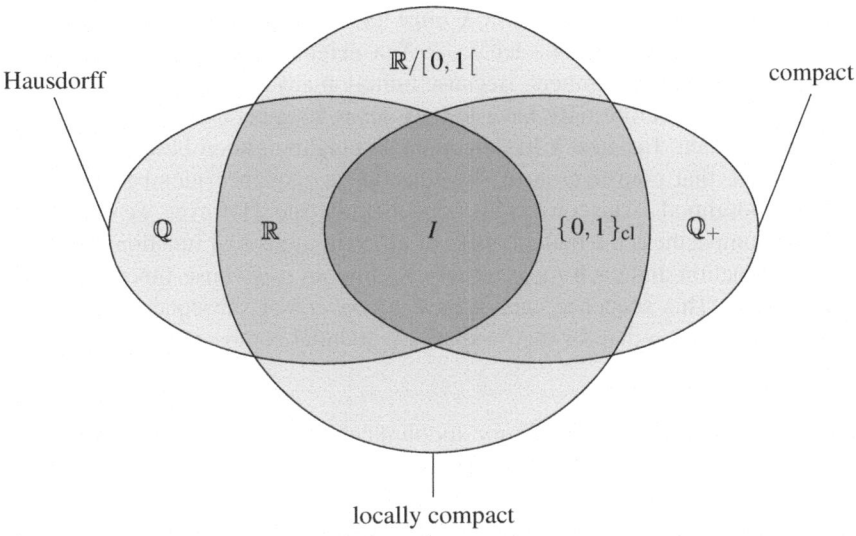

Fig. 4.4 Implications between different concepts of compactness

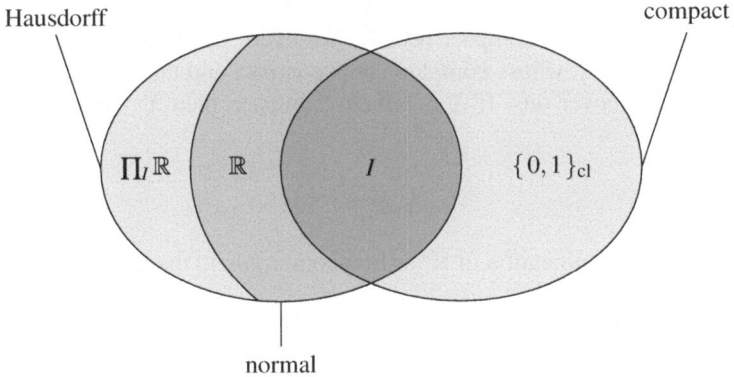

Fig. 4.5 Implications between various separation properties

sequentially compact. For example, the sequence of functions defined by

$$f_n(0, a_1 a_2 \ldots) = a_n$$

that assigns to a decimal number in the interval $[\,0, 1\,[$ its n–th digit, has no convergent subsequence in the product $\prod_{[\,0,1\,[}\{0, \ldots, 9\}$. Such a convergent subsequence would indeed provide for each $a = (0, a_1 a_2 \ldots) \in [\,0, 1\,[$ also a convergent subsequence (a_n) in $\{0, \ldots, 9\}$, which leads to a contradiction. In Sect. 4.3, it is shown that the product is compact. For metric spaces (more generally for spaces that satisfy the first countability axiom), however, compactness implies sequential compactness. Every sequence (a_n) in X must have at least one *accumulation point*, because otherwise there is for each $x \in X$ a neighbourhood that contains only finitely many sequence members. Because finitely many such already cover X, the sequence would therefore only have finitely many members. So let a be such an accumulation point. Because a has a countable neighbourhood base, we can form a subsequence that converges to a, by selecting a sequence member in each of these neighbourhoods. The converse is not generally true. However, we obtain from sequential compactness the boundedness of all continuous real functions on X. For each such function and each n, we otherwise find an a_n, whose function value is greater than n. This sequence cannot have a convergent subsequence. It follows from the last sentence that for metric spaces sequential compactness is the same as compactness.

One-Point Compactification Every topological space X can be extended by adding a new point ∞ to make a compact space

$$X_+ = X \cup \{\infty\}$$

in which X is contained as a subspace. For this purpose, let a subset U of X_+ be open if it lies in X and is open therein or if $\infty \in U$ and $X_+ \setminus U$ is closed in X and compact. This topology makes X_+ compact because for every open cover $(U_j \mid j \in J)$ of X_+ there is a set U_j, whose complement is compact and thus covered by finitely many of the other cover sets. If X is already compact, then X_+ is the topological sum

$$X_+ \cong X + \{\infty\}.$$

The one-point compactification of $\mathbb{R}^n$ is homeomorphic to the sphere S^n.

Exercises

Exercise 64 Unite
Finite unions of compact subspaces are compact.

Exercise 65 The Topological Quadrature of the Circle
Show that the circle, i.e., the topological space $S^1 = \{x \in \mathbb{R}^2 \mid \|x\|_2 = 1\}$, is homeomorphic to the square $\{x \in \mathbb{R}^2 \mid \|x\|_\infty = 1\}$.

Exercise 66 Divide et Impera
Every continuous bijection $\mathbb{R} \to \mathbb{R}$ is a homeomorphism.

Exercise 67 The Whole Zariski
Is $\mathrm{Spec}(\mathbb{Z})$ compact?

Exercise 68 Everything Normal
Every compact Hausdorff space is normal.

Exercise 69 Projection Closed
Prove that a topological space X is compact if and only if the canonical projection $\text{pr}_Y : X \times Y \to Y$ is closed for every space Y.

Exercise 70 Small and Large
A topological product of countably many factors is sequentially compact (see supplement) if and only each factor is. However, the space $\prod_\mathbb{R} S^0$ is not sequentially compact.

Exercise 71 ∞ Is 0
The one-point compactification of the discrete space $\mathbb{N}$ is homeomorphic to the subspace $\{0\} \cup \{1/n \mid n \in \mathbb{N}\} \subseteq \mathbb{R}$. What happens for $\mathbb{Z}$ instead of $\mathbb{N}$?

Exercise 72 Complete
Let X be a metric space. A *Cauchy sequence* in X is known to be a sequence $(x_n \mid n \in \mathbb{N})$ with the property that for every $\varepsilon > 0$ we have

$$d(x_n, x_m) < \varepsilon$$

for sufficiently large n, m. The space X is called *complete*, if every Cauchy sequence converges. Show that X is totally bounded if and only if every sequence contains a subsequence that is a Cauchy sequence. Conclude from this that a metric space is compact if and only if it is totally bounded and complete.

Exercise 73 Cut!
Let K and L be compact subspaces of a topological space. Must the subspace $K \cap L$ be compact, too?

Exercise 74 Only Rational
The space $\mathbb{Q}$ is not locally compact with respect to the usual topology.

Exercise 75 Locally Compact Product
If $(X_i \mid i \in I)$ is a family of topological spaces, and their product $\prod_{i \in I} X_i$ is locally compact, then each individual space X_i is locally compact and almost all X_i are compact.

4.2 Proper Maps

Often, there are properties of continuous maps that correspond to properties of topological spaces in such a way that a space X has the property if and only if the map $X \to \star$ to a singleton space has the corresponding property. This idea was already suggested when we discussed separated maps, which relativise the Hausdorff property (see the exercises in Sect. 58). This section is about a relativisation of the compactness property.

Fig. 4.6 The fibres in the Möbius strip

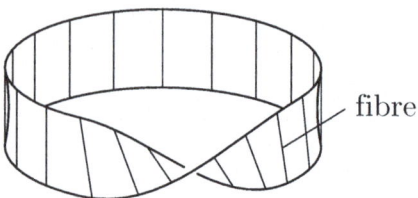

Definition 4.2.1

A continuous map $f: X \to Y$ is called *proper* if it is closed and all pre-images $f^{-1}(y)$ are compact. (Instead of pre-images, we often speak of *fibres*. This is also much easier to read.)

Examples 4.2.2
The projection from the Möbius strip

$$p: ([0,1] \times [0,1])/((0,t) \sim (1, 1-t)) \longrightarrow [0,1]/\partial[0,1]$$

onto the first coordinate is proper. The fibres are closed intervals (see Fig. 4.6).
More generally, every continuous map from a compact space into a Hausdorff space is proper. However, there are also proper maps whose source is not compact. For example, homeomorphisms are always proper.

Theorem 4.2.3
For every proper map, the pre-images of compact sets are compact.

Proof. Let $f: X \to Y$ be proper and $K \subseteq Y$ be compact. It suffices to show that $f^{-1}(K) \subseteq X$ is compact. So, let $(U_j \mid j \in J)$ be an open cover of $f^{-1}(K)$. Then there are open subsets V_j of X with $U_j = f^{-1}(K) \cap V_j$. For each k in K, the fibre $f^{-1}(k)$ is compact by assumption. So there is a finite subset $J(k)$ of J, such that the fibre $f^{-1}(k)$ is contained in

$$V(k) = \bigcup_{j \in J(k)} V_j.$$

This set is open. By assumption, then

$$W(k) = Y \setminus f(X \setminus V(k))$$

is also open. Note that we have $k \in W(k)$ and $f^{-1}(W(k)) \subseteq V(k)$. This provides an open cover $(W(k) \mid k \in K)$ of K. By assumption, there is, therefore, a finite

4.2 Proper Maps

subset K_0 of K with

$$K \subseteq \bigcup_{k \in K_0} W(k).$$

Then we have

$$f^{-1}(K) \subseteq \bigcup_{k \in K_0} f^{-1}(W(k)) \subseteq \bigcup_{k \in K_0} \bigcup_{j \in J(k)} V_j.$$

This provides the desired finite subcover. $\square$

The converse does not necessarily hold. For every finite set, the identity is a continuous map from the discrete topology to the clump topology. All pre-images are finite and thus compact. However, the map is not closed if the set has more than one element.

Corollary 4.2.4
The composition of proper maps is proper.

Proof. This is true because pre-images of compact sets under proper maps are compact. $\square$

Topological spaces, together with proper maps, thus form a subcategory of the usual category of topological spaces and (all) continuous maps.

The following theorem about proper maps has numerous consequences.

Theorem 4.2.5
Every base change of every proper map is proper.

Proof. Let $p \colon X \to B$ be proper and

$$\begin{array}{ccc} X' & \longrightarrow & X \\ {\scriptstyle p'} \downarrow & & \downarrow {\scriptstyle p} \\ B' & \xrightarrow{f} & B \end{array}$$

a pullback with $X' = B' \times_B X$. The fibres of p' coincide with those of p, because

$$p'^{-1}(b') = \{(b', x) \mid f(b') = p(x)\} \cong p^{-1}(f(b')).$$

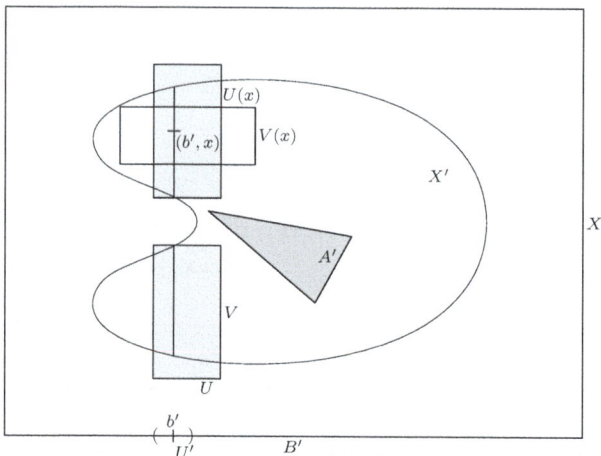

Fig. 4.7 For the proof of Theorem 4.2.5

They are, therefore, compact by assumption. It remains to be shown that p' is also closed. For this, let A' be closed in X' and b' a point from $B' \setminus p'(A')$. If x is in the fibre of p over $f(b')$, then (b', x) is in the fibre of p' over b' and thus not in A'. There are therefore open neighbourhoods $U(x)$ of b' in B' and $V(x)$ of x in X, so that $U(x) \times V(x)$ is disjoint from A' (see Fig. 4.7). Because the fibre over $f(b')$ is compact, a finite number of $V(x)$ is sufficient to cover this fibre. Let V be their union and U the intersection of the corresponding $U(x)$. Then also $U \times V$ and A' are disjoint. And the subset $W = B \setminus p(X \setminus V)$ of B is an open neighbourhood of $f(b')$, because by construction V contains the fibre of p over $f(b')$. Due to the continuity of f there is therefore an open neighbourhood $U' \subseteq U$ of b' with $f(U') \subseteq W$. Then U' is also disjoint from $p'(A')$. Otherwise, there would be a point (u', x) in A' with u' from U', so $f(u') = p(x)$ from W. This would imply that x is in V, a contradiction to $(U \times V) \cap A' = \emptyset$. □

Corollary 4.2.6 (Product Theorem)
If $f: X \to Y$ and $f': X' \to Y'$ are proper, then their product

$$f \times f': X \times X' \to Y \times Y'$$

is also proper. In particular, if X and X' are compact, then their product $X \times X'$ is also compact.

4.2 Proper Maps

Proof. Since $f \times f'$ is the composition $(f \times \mathrm{id}_{Y'})(\mathrm{id}_X \times f')$, we can assume that one of the two maps is the identity. But $f \times \mathrm{id}_{Y'}$ is up to canonical homeomorphism the base change of f along pr_Y

$$\begin{array}{ccc} X \times Y' & \longrightarrow & X \\ {\scriptstyle f \times \mathrm{id}_{Y'}} \downarrow & & \downarrow {\scriptstyle f} \\ Y \times Y' & \xrightarrow{\mathrm{pr}_Y} & Y \end{array}$$

and thus proper. □

As a simple corollary, we obtain:

Corollary 4.2.7 (Heine–Borel)
A subspace of $\mathbb{R}^n$ is compact if and only if it is closed and bounded.

Proof. Since $\mathbb{R}^n$ is a Hausdorff space, every compact subspace must be closed. In addition, the space $\mathbb{R}^n$ is covered by the open ε–neighbourhoods of the origin, if we exceptionally also think of the large ε. Every compact subspace lies in finitely many, thus in the one with the largest ε. Hence, compact sets are also bounded. Conversely, if a subset is bounded, it lies in a large cube. This is homeomorphic to a n–fold product of compact intervals, thus itself compact. If the subset within it is also closed, it is automatically compact. □

Exercises

Exercise 76 Compactified Maps
If $f: X \to Y$ is proper, then the map

$$f_+ : X_+ \longrightarrow Y_+$$

that satisfies $f_+(x) = f(x)$ for $x \in X$ and $f(\infty) = \infty$, is continuous. (The one-point compactification X_+ was defined in Sect. 4.1).

Exercise 77 The Cantor's Wipe Set
Let W be the set of all $t \in [\,0, 1\,]$ with

$$3^n t - \lfloor 3^n t \rfloor \notin \,]\,1/3, 2/3\,[$$

for all $n \in \mathbb{N}$. (Here, the Gauss bracket $\lfloor x \rfloor$ for every $x \in \mathbb{R}$ denotes the largest integer m with $m \leqslant x$.) Show that W is compact and thus homeomorphic to $\prod_{\mathbb{N}} S^0$.

4.3 Tychonoff's Theorem

This section is dedicated to the following theorem about the compactness of arbitrary products.

Theorem 4.3.1 (Tychonoff)
For any index sets I, the product $\prod_{i \in I} X_i$ of compact spaces X_i is compact.

The statement is false if compactness is replaced by sequential compactness (see Sect. 4.1). The reason is, somewhat vaguely speaking, because arbitrary products of countable spaces do not need to be countable again; even the space $\{0, 1\}^{\mathbb{N}}$ is uncountable. So, here we see another reason for not characterising topological spaces by properties that are defined in terms of sequences. With the cover property, the theorem is correct, but new techniques are required for proof. A proof (using filters) is given in this section. It can be skipped on the first reading. First, we introduce some terms that further characterise the compactness property. This characterisation can then be transferred from the factors to the product.

Definition 4.3.2

Let X be a set. A *filter* on X is a subset $\mathcal{F}$ of the power set of X with the following properties

(F1) $F \in \mathcal{F}, F \subseteq F' \subseteq X \implies F' \in \mathcal{F}$
(F2) $F_1, F_2 \in \mathcal{F} \implies (F_1 \cap F_2) \in \mathcal{F}$
(F3) $\mathcal{F} \neq \emptyset$
(F4) $\emptyset \notin \mathcal{F}$

It is best to imagine that a filter only contains objects of a certain 'size' that it 'filters out'.

Example 4.3.3
For example, for every point x of a topological space X the set $\mathcal{U}(x)$ of neighbourhoods is a filter on the set X, the *neighbourhood filter* of x.

Definition 4.3.4

A *filter base* $\mathcal{B}$ on X is a set of subsets of X with

(B1) $B_1, B_2 \in \mathcal{B} \implies$ There is a $B \in \mathcal{B}$ with $B \subseteq B_1 \cap B_2$.
(B2) $\mathcal{B} \neq \emptyset$
(B3) $\emptyset \notin \mathcal{B}$

4.3 Tychonoff's Theorem

A filter base $\mathcal{B}$ generates a filter.

$$\langle \mathcal{B} \rangle = \{F \subseteq X \mid \text{There is a } B \text{ in } \mathcal{B} \text{ with } B \subseteq F.\}$$

For example, in metric spaces, the neighbourhood filters are countably generated:

$$\langle \{U_{1/n}(x) \mid n \in \mathbb{N}\} \rangle = \mathcal{U}(x).$$

For topological spaces X, we can introduce a concept of convergence for filters as follows.

Definition 4.3.5

A filter $\mathcal{F}$ *converges* to $x \in X$ if $\mathcal{F}$ contains the neighbourhood filter $\mathcal{U}(x)$.

This notion of convergence is closely related to the ordinary concept of convergence of sequences $(x_n \mid n \in \mathbb{N})$. If $\mathcal{F}$ is the filter of all subsets that contain almost all sequence members, then the sequence converges to x if and only if the filter $\mathcal{F}$ converges to x.

Definition 4.3.6

A filter $\mathcal{F}$ on X is called an *ultrafilter* if every other filter on X that contains $\mathcal{F}$ coincides with $\mathcal{F}$.

Zorn's lemma from set theory states that in partially ordered sets, there are always maximal elements as long as all ordered subsets have upper bounds. When we apply this to the set of all filters that contain $\mathcal{F}$, we obtain the following result.

Theorem 4.3.7
Every filter is contained in an ultrafilter.

Proof. The upper bounds are obtained here by the unions of the filters. □

Ultrafilters have a surprising property.

Theorem 4.3.8
Let $\mathcal{F}$ be an ultrafilter on X and A a subset of X. Then $\mathcal{F}$ always contains A itself or the complement of A.

Proof. If there is a $F \in \mathcal{F}$ that does not intersect A, then the complement of A includes the set F and is therefore in the filter according to (F1). On the other hand, if all sets of the filter intersect A, then

$$\mathcal{F}' = \langle \{F \cap A \mid F \in \mathcal{F}\} \rangle$$

is a filter that contains A and the sets of $\mathcal{F}$ and therefore coincides with $\mathcal{F}$. Thus, the set A is then in the filter. □

Theorem 4.3.9
A topological space is compact if and only if every ultrafilter on it converges.

Proof. Suppose $\mathcal{F}$ is a non-convergent ultrafilter on X. Then for every $x \in X$ there is an open neighbourhood $U(x)$ that is not in the filter. If X is compact, then finitely many $U(x_1), U(x_2), \ldots, U(x_n)$ cover X. Because these sets are not in the ultrafilter, their complements must be in $\mathcal{F}$. However, since their intersection is empty, this contradicts (F4). Conversely, if $(U_i \mid i \in I)$ is a cover of X without a finite subcover, then

$$\langle \{X \setminus \bigcup_{i \in E} U_i \mid E \subseteq I \text{ finite }\} \rangle$$

defines a filter. Let $\mathcal{F}$ be an ultrafilter that contains this filter, and x its limit. The element x is contained in a set U_{i_0} and thus U_{i_0} is in the filter $\mathcal{F}$. However, the filter cannot contain both U_{i_0} and its complement simultaneously because their intersection is empty. □

Proof of Tychonoff's Theorem. To prove Tychonoff's theorem, it is now sufficient to verify the convergence of ultrafilters in the product $X = \prod_i X_i$. If $\mathcal{F}$ is an ultrafilter, then every set generated by its projections

$$\mathcal{F}_i = \{\mathrm{pr}_i F \mid F \in \mathcal{F}\}$$

is an ultrafilter on X_i. For example, the filter property (F1) of $\mathcal{F}_i$ can be seen as follows. Let U be a superset of $\mathrm{pr}_i F$ for a $F \in \mathcal{F}$. Then the set $\mathrm{pr}_i^{-1} U$ is also in $\mathcal{F}$, because it contains F and thus

$$U = \mathrm{pr}_i(\mathrm{pr}_i^{-1} U) \in \mathcal{F}_i.$$

To prove the ultrafilter property of $\mathcal{F}_i$, let $\mathcal{F}_i$ be contained in a filter $\mathcal{G}$ on X_i. Then it defines a filter base

$$\mathcal{B} = \{(\mathrm{pr}_i^{-1} G) \cap F \mid G \in \mathcal{G}, F \in \mathcal{F}\}$$

4.3 Tychonoff's Theorem

that contains $\mathcal{F}$. So it generates $\mathcal{F}$, because $\mathcal{F}$ is an ultrafilter. For each $G \in \mathcal{G}$, therefore, the set $\text{pr}_i^{-1} G$ is in $\mathcal{F}$ and thus $G = \text{pr}_i(\text{pr}_i^{-1} G)$ is in $\mathcal{F}_i$. Due to the compactness of the X_i, the filter $\mathcal{F}_i$ converges to an $x_i \in X_i$. Set

$$x = (x_i \mid i \in I).$$

Then $\mathcal{F}$ converges to x, because if U is a neighbourhood of x in the product, there are neighbourhoods U_1 of x_{i_1} in $X_{i_1}, \ldots, U_k$ of x_{i_k} in X_{i_k} with

$$U \supseteq (\text{pr}_{i_1}^{-1} U_1 \cap \ldots \cap \text{pr}_{i_k}^{-1} U_k).$$

The U_j are in the neighbourhood filter and thus in $\mathcal{F}_{i_j}$ for all j. So there is $V_j \in \mathcal{F}$ with

$$U_j = \text{pr}_{i_j} V_j,$$

and thus the supersets $\text{pr}_{i_j}^{-1} U_j$ of V_j are in $\mathcal{F}$. Because filters are closed with respect to finite intersections and supersets, the set U is also in the filter, and the proof of convergence is complete. □

Supplement

Stone–Čech Compactification Besides the one-point compactification, there is another compactification that has particularly good properties for completely regular spaces X. To describe this, let F be the set of all continuous functions from X to the interval $I = [0, 1]$ and

$$i : X \longrightarrow \prod_F I$$

given by $i(x) = (f(x) \mid f \in F)$. Then according to Tychonoff's theorem,

$$\widetilde{X} = \overline{i(X)} \subseteq \prod_F I$$

is compact. This is the *Stone–Čech compactification* of X. It can be shown that a continuous map from X into a compactum can always be uniquely extended to $\widetilde{X}$.

4.4 Mapping Spaces

In linear algebra, it is an essential insight that the set of linear maps between two vector spaces V and W also carries the structure of a vector space. This vector space is usually denoted by $\mathrm{Hom}(V, W)$. There is something comparable in topology as well. The set of continuous maps between two topological spaces X and Y can itself be equipped with a topological structure. A similar approach is known from functional analysis, where the function spaces are also equipped with topologies. In contrast to functional analysis, the more general framework of topology is much larger, and only in exceptional cases will we recognise the familiar topologies here (see, for example, Theorem 4.4.4). The topology we introduce on mapping spaces here is, however, not unrestrictedly useful, so the theorems listed below always have some technical prerequisites. One of these prerequisites is the local version of the compactness property. Alternatively, we can restrict the consideration to compactly generated spaces. This approach will be detailed at the end of this chapter in Sect. 4.5.

Let X and Y be topological spaces. We will now equip the set $\mathrm{Hom}(X, Y)$ of continuous maps from X to Y with a topology. If $K \subseteq X$ is compact and $V \subseteq Y$ is open, then at least

$$M(K, V) = \{ f \in \mathrm{Hom}(X, Y) \mid f(K) \subseteq V \}$$

should be open.

Definition 4.4.1

A subset $U \subseteq \mathrm{Hom}(X, Y)$ is open in the *compact-open topology* if, for each map f that lies in U, there are finitely many K_j and V_j with

$$f \in \bigcap_j M(K_j, V_j) \subseteq U.$$

In other words, the sets $M(K, V)$ form the subbasis of the compact-open topology on $\mathrm{Hom}(X, Y)$.

Example 4.4.2
To become familiar with this definition, let us first assume that the topological space $X = \{1, \ldots, n\}$ is actually a finite discrete set. Then

$$\varphi \colon \mathrm{Hom}(\{1, \ldots, n\}, Y) \longrightarrow Y^n, \ f \longmapsto (f(1), \ldots, f(n))$$

is continuous because

$$\varphi^{-1}(V_1 \times \cdots \times V_n) = \bigcap_{j=1}^n M(\{j\}, V_j).$$

4.4 Mapping Spaces

Also, the inverse map

$$\psi: Y^n \longrightarrow \mathrm{Hom}(\{1,\ldots,n\},Y),\ (y_1,\ldots,y_n) \longmapsto (j \mapsto y_j)$$

is continuous because

$$\psi^{-1}M(K,V) = \bigcap_{j \in K} \mathrm{pr}_j^{-1}(V)$$

is always open in Y^n. Thus, in this special situation, we have shown that there is homeomorphism $\mathrm{Hom}(\{1,\ldots,n\},Y) \cong Y^n$.

Example 4.4.3
The space $\mathrm{Hom}(I,I)$ of continuous self-maps of the unit interval is not compact. To see this from the definition, we follow [Gro12] and recall from the intermediate value theorem that every continuous map $I \to I$ has a fixed point. This means that, for any $\varepsilon > 0$, the open subsets

$$U(x) = M(\{x\},\ I \cap]x-\varepsilon, x+\varepsilon[\ )$$

cover the space $\mathrm{Hom}(I,I)$. We will show that there is no finite subcover of this particular cover. For that purpose, let us look at finitely many of them, say $U(x_1),\ldots,U(x_n)$ with $x_1 < \cdots < x_n$. If $\varepsilon < 1/2$, which we can now assume too, then $I \cap]x-\varepsilon, x+\varepsilon[\neq I$ for any x. Thus, we can find, for each index j, a point $y_j \notin I \cap]x_j - \varepsilon, x_j + \varepsilon[$. Then we can find a continuous function $f: I \to I$ that satisfies $f(x_j) = y_j$ for all j. By construction, this function f is not in any open subset $U(x_j)$. We deduce that there is no finite subcover for this particular cover.

The following theorem states that the compact-open topology provides something familiar in situations typical for analysis.

> **Theorem 4.4.4**
> *Let X be a compact Hausdorff space and Y a metric space. Then the compact-open topology on $\mathrm{Hom}(X,Y)$ is induced by the metric*
>
> $$d(f,g) = \sup_{x \in X} d(f(x), g(x))$$
>
> *of uniform convergence.*

Proof. We first consider a map f from a compact-open open set of the form $M(K,V)$. We want to show that this set also contains an ε–neighbourhood of f. In any case, we have $f(K) \subseteq V$. For each point y from $f(K)$ there is an $\varepsilon(y)$, so that the $\varepsilon(y)$–neighbourhood of y is entirely in V. The $\varepsilon(y)/2$–neighbourhoods of the y cover all of $f(K)$. So there is a finite subset $Y_0 \subseteq f(K)$, so that the $\varepsilon(y)/2$–neighbourhoods of the y from Y_0 cover all of $f(K)$. Let ε be the minimum of the $\varepsilon(y)$ with y in Y_0. If then a map g is in the $\varepsilon/2$–neighbourhood of f and x from K, then $g(x)$ is in the $\varepsilon/2$–neighbourhood of $f(x)$, and $f(x)$ is in the $\varepsilon(y)/2$–neighbourhood of a y from Y_0. Because $\varepsilon \leqslant \varepsilon(y)$, the point $g(x)$ is then in the $\varepsilon(y)$–neighbourhood of y, thus in V. This shows $g(K) \subseteq V$. The $\varepsilon/2$–neighbourhood of f is therefore entirely in $M(K, V)$. Using the minimum, we see that even the finite intersections of sets of the form $M(K, V)$ still contain ε–neighbourhoods of their points.

Let now f and ε be given. For each point x from X let $V(x)$ be the $\varepsilon/2$–neighbourhood of $f(x)$. Their pre-image is an open neighbourhood of x. Thus, it contains a compact neighbourhood $K(x)$. There is then a finite set $X_0 \subseteq X$ with

$$X = \bigcup_{x \in X_0} K(x).$$

By construction, the map f lies in the compact-open open set

$$\bigcap_{x \in X_0} M(K(x), V(x)).$$

If now g lies therein, then g lies in the ε–neighbourhood of f. Each x' from X lies in a $K(x)$, and then $f(x')$ and $g(x')$ lie in $V(x)$, thus have a distance to $f(x)$ smaller than $\varepsilon/2$. Their distance to each other is therefore smaller than ε. □

After clarifying how the compact-open topology looks in relevant examples we can now study it systematically. This means in particular, that we do not simply construct a new space $\mathrm{Hom}(X, Y)$ from spaces X and Y; we should also immediately consider the compatibility of this construction with continuous maps.

Theorem 4.4.5

For every continuous map $f: Y \to Y'$, the assignment $f_*(g) = fg$ induces a continuous map

$$f_*: \mathrm{Hom}(X, Y) \longrightarrow \mathrm{Hom}(X, Y'),$$

and for every continuous map $f: X \to X'$, the assignment $f^*(g) = gf$ induces a continuous map

$$f^*: \mathrm{Hom}(X', Y) \longrightarrow \mathrm{Hom}(X, Y).$$

Proof. From

$$(f_*)^{-1} M(K, V') = M(K, f^{-1}(V'))$$

4.4 Mapping Spaces

and

$$(f^*)^{-1} M(K, V) = M(f(K), V)$$

it immediately follows that the pre-images of open sets are again open. □

Let $f: X \times Y \to Z$ be a continuous map. Then for each x from X, the map

$$f^\#(x): Y \longrightarrow Z, y \longmapsto f(x, y)$$

is continuous. This, therefore, provides a map

$$f^\#: X \longrightarrow \mathrm{Hom}(Y, Z).$$

Definition 4.4.6

The map $f^\#$ is called the *adjoint* to the map f.

Theorem 4.4.7
For every continuous map, the adjoint map is also continuous.

Proof. It suffices to show that the pre-images of the sets $M(L, W)$ are open when $L \subseteq Y$ is compact and $W \subseteq Z$ is open. Let x be a point of the pre-image. Then $f(\{x\} \times L) \subseteq W$. Due to the continuity of f and the compactness of L, there is even an open neighbourhood U of x in X with $f(U \times L) \subseteq W$ (see Fig. 4.8). But that means that all of U lies in the pre-image. The pre-image is, therefore, open. □

The preceding theorem shows the most important method for defining continuous maps into mapping spaces.

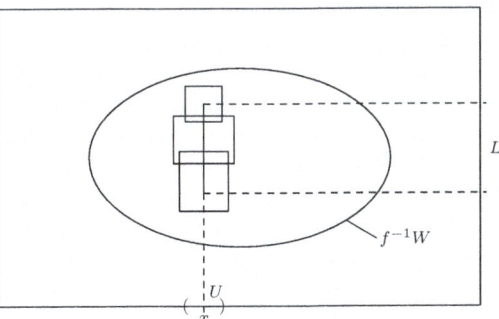

Fig. 4.8 For the proof of Theorem 4.4.7

Adjoining provides a map

$$\mathrm{Hom}(X \times Y, Z) \longrightarrow \mathrm{Hom}(X, \mathrm{Hom}(Y, Z)).$$

It is always injective, and we can ask under what conditions it is also surjective, thus bijective.

Theorem 4.4.8 (Exponential Law)
If Y is locally compact, then the adjunction

$$\mathrm{Hom}(X \times Y, Z) \longrightarrow \mathrm{Hom}(X, \mathrm{Hom}(Y, Z))$$

is bijective.

Proof. Only the surjectivity remains to be shown. Let $f \colon X \times Y \to Z$ be a map whose adjoint map $f^\# \colon X \to \mathrm{Hom}(Y, Z)$ is continuous. Then it suffices to show the continuity of f at each point (x, y). Because $f^\#(x)$ is continuous and Y is locally compact, there is for every open neighbourhood W of $f(x, y)$ in Z a compact neighbourhood L of y in Y with $f^\#(x)(L) \subseteq W$. Since $f^\#$ is continuous, the set

$$U = (f^\#)^{-1} M(L, W) = \{x' \in X \mid f^\#(x')(L) \subseteq W\}$$

is an open neighbourhood of x in X. But then $U \times L$ is a neighbourhood of (x, y) with $f(U \times L) = f^\#(U)(L) \subseteq W$. □

The source and target of the adjunction $f \mapsto f^\#$ also carry the compact-open topology, and we might wonder whether this map is continuous, open, or even a homeomorphism. This will not be pursued at this point.

Corollary 4.4.9
If X is locally compact, then the evaluation

$$\mathrm{ev} \colon \mathrm{Hom}(X, Y) \times X \longrightarrow Y, \ (f, x) \longmapsto f(x)$$

is continuous.

Proof. The evaluation is adjoint to the identity of $\mathrm{Hom}(X, Y)$. □

4.4 Mapping Spaces

Corollary 4.4.10
If X and Y are locally compact, then the composition

$$\text{Hom}(Y, Z) \times \text{Hom}(X, Y) \longrightarrow \text{Hom}(X, Z), \quad (g, f) \longmapsto gf$$

is continuous.

Proof. It is sufficient in each case to consider the adjoint map. The diagram

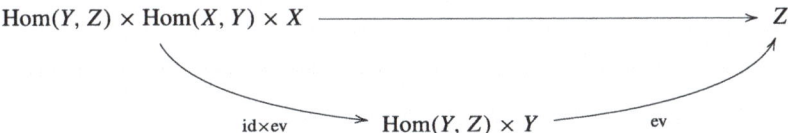

shows that the adjoint map is the composition of two continuous maps. □

To conclude this chapter, we will demonstrate an elegant application of the exponential law, in which the universal properties of all constructions come into play.

Corollary 4.4.11
Given a locally compact space Y and an identification $p: X \to X'$, the product

$$p \times \text{id}: X \times Y \longrightarrow X' \times Y$$

is also an identification.

Proof. Let T be a topological space. Let $f: X' \times Y \to T$ be a map and $f(p \times \text{id})$ be continuous. It remains to show that then f is also continuous. In the diagram

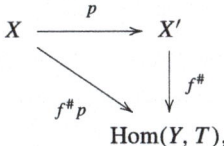
$$\text{Hom}(Y, T),$$

the composition $f^\# p$ is continuous because it is adjoint to $f(p \times \text{id})$. Therefore, it follows that $f^\#$ is continuous because p is an identification. Finally, the continuity for f is again derived from the exponential law. □

Exercises

Exercise 78 Constant Maps
If X and Y are topological spaces, each point from Y defines the corresponding constant map:
$$Y \longrightarrow \mathrm{Hom}(X, Y), \quad y \longmapsto (x \mapsto y).$$
This map is continuous. What does the adjoint look like?

Exercise 79 Hausdorff Mapping Spaces
A topological space Y is a Hausdorff space if and only if for all topological spaces X the mapping space $\mathrm{Hom}(X, Y)$ is a Hausdorff space.

Exercise 80 Product and Sum
Let X, Y and Z be topological spaces. The universal properties of the product and the sum provide bijections
$$\mathrm{Hom}(X + Y, Z) \xrightarrow{\cong} \mathrm{Hom}(X, Z) \times \mathrm{Hom}(Y, Z),$$
$$\mathrm{Hom}(X, Y \times Z) \xrightarrow{\cong} \mathrm{Hom}(X, Y) \times \mathrm{Hom}(X, Z).$$
Are these continuous? Are they homeomorphisms?

Exercise 81 Exponential Missing
Let P be constructed from $\mathbb{R}$ by identifying the subset $\mathbb{Z}$ to a point, and denote by $p \colon \mathbb{R} \to P$ the quotient map. Then the map
$$p \times \mathrm{id} \colon \mathbb{R} \times \mathbb{Q} \longrightarrow P \times \mathbb{Q}$$
is neither closed nor an identification.

4.5 Locally-Compactly Generated Spaces

In this section, which can be skipped at the first reading, a category of topological spaces is introduced in which the exponential laws apply without restriction and which is equipped with the usual constructions such as subspaces, products, sums and identifications. The primary reference is [McC69]. However, unlike in the original treatment, the Hausdorff property of the compact spaces is not required, and instead, we work with locally compact spaces. This approach is appropriate because a general form of the exponential laws for all locally compact spaces has already been proven. The presentation becomes clearer, and fewer spaces need to be re-topologised. The classical compactly generated spaces are obtained when the locally compact spaces are replaced by compact Hausdorff spaces.

4.5 Locally-Compactly Generated Spaces

Definition 4.5.1

Let $\mathcal{K}$ be the class of locally compact spaces. For a topological space X, we set

$$X(\mathcal{K}) = \{s \colon K \to X \mid s \text{ is continuous and } K \in \mathcal{K}\}.$$

A topological space X is called *locally-compactly generated* or shorter k–space, if it carries the co-induced topology from $X(\mathcal{K})$.

In other words, a map $f \colon X \to T$ into any space T is continuous if and only if for every continuous map $s \colon K \to X$, the composition fs is continuous.

Examples 4.5.2

All locally compact spaces are locally-compactly generated, as we quickly realise. If X satisfies the first countability axiom (for example, because X is a metric space), then X is a k–space. To see this, let $A \subseteq X$ be a subset of X for which $s^{-1}A$ is closed for all $s \in X(\mathcal{K})$. Let further $a \in \overline{A}$ and $(U_n \mid n \in \mathbb{N})$ be a neighbourhood base of a. Construct a convergent sequence (a_n) with

$$a_n \in A \cap U_1 \cap U_2 \cap \ldots \cap U_n.$$

Then the map

$$s \colon \{0\} \cup \{1/n \mid n \in \mathbb{N}\} \longrightarrow X$$

with $s(n) = a_n$ and $s(0) = a$ is continuous. The pre-image $s^{-1}A$ contains the sequence $(1/n)$ and is closed by assumption. So, the point 0 lies in the pre-image, and we have $a \in A$.

If X is any topological space, it can always be made into a locally-compactly generated space by changing the topology as follows:

Definition 4.5.3

For a topological space $(X, \mathcal{T})$ denote kX the space $(X, k\mathcal{T})$, where $k\mathcal{T}$ is the co-induced topology by $X(\mathcal{K})$. We call kX the *locally-compactly generated space* of X.

▶ **Remark 4.5.4** If $f \colon X \to Y$ is continuous, then so is the map

$$kf \colon kX \longrightarrow kY, \; x \longmapsto f(x).$$

Note that kX coincides with X if X is already locally-compactly generated. The following theorem states that the full subcategory of locally-compactly generated spaces is closed under all the important universal constructions. When co-inducing, the spaces remain unchanged.

Theorem 4.5.5
(a) If $p\colon X \to X'$ is an identification, then with X also X' is locally-compactly generated.
(b) If X_j is locally-compactly generated for each $j \in J$, then so is their sum $\coprod_{j \in J} X_j$.
(c) If X is locally-compactly generated and $i\colon M \subseteq X$ is the inclusion of a subspace, then $k(i)\colon kM \to X$ has the universal property of a subspace for all k–spaces.
(d) If X_j is locally-compactly generated for each $j \in J$, then $k(\prod_{j \in J} X_j)$ has the universal property of a product for all k–spaces.

Proof. Let $f\colon X' \to T$ be a map, whose composition with all $s'\colon K' \to X'$ is continuous. Then this applies in particular to all s' of the form ps with $s \in X(\mathcal{K})$. Because X is locally-compactly generated, the map fp must therefore be continuous. Finally, the map f must be continuous, because p is an identification. The statement (b) is shown in a completely analogous way. For (c) let $f\colon T \to k(M)$, the space T be locally-compactly generated and $k(i)f$ be given as continuous. We must show that f is continuous. For this, it is sufficient to prove the continuity of fs for each $s\colon K \to T$. In the diagram

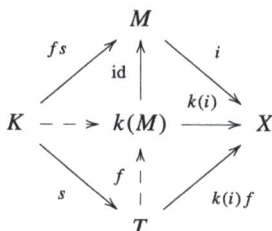

the map $fs\colon K \to M$ is continuous, because M has the subspace property and $k(i)fs$ is continuous. But this also makes the map from K to $k(M)$ continuous because K is locally compact and an open subset of $k(M)$ always has an open preimage among all continuous maps of a locally compact space to M. We proceed similarly with (d). □

Definition 4.5.6

For k–spaces X and Y, we set

$$X \times_k Y = k(X \times Y)$$

4.5 Locally-Compactly Generated Spaces

and
$$M \subseteq_k X = k(M).$$

Example 4.5.7

If X is locally-compactly generated and Y is locally compact, then the product $X \times Y$ is locally-compactly generated, and we have
$$X \times_k Y = X \times Y.$$

To see this, let $f : X \times Y \to T$ be any map that is continuous when combined with any continuous map from a locally compact space. It suffices to prove the continuity of the map adjoint to f, denoted
$$f^{\#} : X \longrightarrow \operatorname{Hom}(Y, T).$$

Because X is locally compact, this can be combined with a $s \in X(\mathcal{K})$. The adjoint to $f^{\#}s$ has the form
$$K \times Y \xrightarrow{s \times \mathrm{id}} X \times Y \xrightarrow{f} T$$

and is continuous because Y is locally compact. Therefore, the assertion follows again from the exponential law.

Definition 4.5.8

Let $\mathcal{T}$ be the topology induced by all maps
$$s^* : \operatorname{Mor}(X, Y) \longrightarrow \operatorname{Hom}(K, Y)$$

for $s \in X(\mathcal{K})$ on $\operatorname{Mor}(X, Y)$, where the targets $\operatorname{Hom}(K, Y)$ carry the compact-open topology. Since the maps s^* are continuous, when $\operatorname{Mor}(X, Y)$ is also equipped with the compact-open topology, the topology $\mathcal{T}$ is possibly coarser than the compact-open topology. The k–mapping spaces are now defined by
$$\operatorname{Hom}_k(X, Y) = (\operatorname{Mor}(X, Y), k\mathcal{T}).$$

Theorem 4.5.9 (Exponential Law for Locally-Compactly Generated Spaces)

For all locally-compactly generated spaces X, Y, and Z, the adjunction map
$$\operatorname{Hom}_k(X \times_k Y, Z) \longrightarrow \operatorname{Hom}_k(X, \operatorname{Hom}_k(Y, Z)), \quad f \mapsto f^{\#}$$

is a homeomorphism.

Proof. First, it must be shown that with f also $f^\#$ is continuous. For this, let $s \in X(\mathcal{K})$. The composition

$$f^\# s \colon K \longrightarrow \mathrm{Hom}_k(Y, Z)$$

is continuous if and only if the same map to $(\mathrm{Mor}(Y, Z), \mathcal{T})$ is continuous, because K is locally compact. The latter is continuous when its composition with every t^* is continuous, where $t \in Y(\mathcal{K})$. However, these maps are adjoint to $f(s \times t)$ and thus continuous. Now assume $f^\#$ to be continuous. Let $s \colon K \to X \times_k Y$ be a continuous map from a local compactum K and let $s_X = \mathrm{pr}_X s$ and $s_Y = \mathrm{pr}_Y s$ be the respective components. Then it suffices to show that the map

$$g = f(s_X \times s_Y) \colon K \times K \longrightarrow Z$$

is continuous because we have $g\Delta = fs$, and the diagonal map $\Delta \colon K \to K \times K$ is continuous. The adjoint to g has the form

$$g^\# = s_Y^* f^\# s_X$$

and is thus continuous. Then the adjunction map is well-defined and bijective. In particular, the evaluation map

$$\mathrm{ev} \colon \mathrm{Hom}_k(X, Y) \times_k X \longrightarrow Y$$

is continuous because $\mathrm{ev}^\# = \mathrm{id}_{\mathrm{Hom}_k(X,Y)}$. The continuity of the adjunction map now results from adjoining the composition

$$\mathrm{Hom}_k(X \times_k Y, Z) \times_k X \times_k Y \xrightarrow{\mathrm{ev}} Z \ .$$

Similarly, the continuity of the inverse is obtained by

$$\mathrm{Hom}_k(X, \mathrm{Hom}_k(Y, Z)) \times_k X \times_k Y \xrightarrow{\mathrm{ev}} \mathrm{Hom}_k(Y, Z) \times_k Y \xrightarrow{\mathrm{ev}} Z \ .$$

Definition 4.5.10

A k–space X is called a *weak Hausdorff space* if the diagonal

$$\Delta = \{(x, x) \mid x \in X\} \subseteq X \times_k X$$

is closed.

4.5 Locally-Compactly Generated Spaces

Example 4.5.11
If X is locally compact, then X is a Hausdorff space if and only if X is a weak Hausdorff space. In general, however, there are more closed sets in the locally-compactly generated product, and the above property is weaker.

The most important properties of weak Hausdorff spaces are summarised in the following omnibus theorem. This class of topological spaces is thus closed towards all important constructions and offers a convenient working environment in many situations of topology.

> **Theorem 4.5.12**
> (a) *Every weak Hausdorff space fulfils the separation property* (T1).
> (b) *If Y is a weak Hausdorff space, so is* $\mathrm{Hom}_k(X, Y)$ *for all X.*
> (c) *All k–products and k–subspaces maintain the weak Hausdorff property.*
> (d) *The exponential laws*
>
> $$\mathrm{Hom}_k(X \times_k Y, Z) \cong \mathrm{Hom}_k(X, \mathrm{Hom}_k(Y, Z))$$
>
> *apply for weak Hausdorff spaces.*
> (e) *Arbitrary sums of weak Hausdorff spaces are weak Hausdorff spaces.*

Proof. Suppose two different points in X cannot be separated by neighbourhoods. Then there is an injective continuous map f of a non-discrete space Z with two points to X. The pre-image of the diagonal Δ_X under $f \times_k f$ is the diagonal of Z, which is therefore closed. This is, however, impossible because Z is not a Hausdorff space. The proof of (b) results from the description of the diagonal of $\mathrm{Hom}_k(X, Y)$ as the intersection

$$\Delta_{\mathrm{Hom}_k(X,Y)} = \bigcap_{x \in X} (\mathrm{ev}_x \times_k \mathrm{ev}_x)^{-1} \Delta_Y$$

of closed sets. For (c), let i be the inclusion map of a k–subspace M of X. Then Δ_M is closed as the pre-image of the diagonal of X under $i \times_k i$. For products, note that the diagonal is the intersection of the diagonal pre-images under the projection products $(\mathrm{pr}_i \times_k \mathrm{pr}_i)$. The exponential laws (d) now result directly from the laws for k–spaces due to (b) and (c). The statement (e) about sums results from the isomorphism

$$(\coprod_{i \in I} X_i) \times_k (\coprod_{i \in I} X_i) \cong \coprod_{i,j \in I} X_i \times_k X_j,$$

that is obtained by checking the universal property from the exponential law. □

> **Theorem 4.5.13**
> *There is a functor (see Sect. 6.2) that assigns a weak Hausdorff space $wH(X)$ to each k–space X, togher with a natural transformation $p \colon X \to wH(X)$, such that the induced map*
>
> $$\mathrm{Mor}(wH(X), Y) \cong \mathrm{Mor}(X, Y)$$
>
> *is bijective for all weak Hausdorff spaces Y. If $q \colon X \to X'$ is an identification and X is a weak Hausdorff space, then*
>
> $$X \xrightarrow{\cong} wH(X) \xrightarrow{wH(q)} wH(X')$$
>
> *has the universal property of an identification in the full subcategory of weak Hausdorff spaces.*

Proof. Let R be the intersection of all equivalence relations on X that are closed in $X \times_k X$. Let $wH(X)$ be the corresponding quotient space and p the projection map. The set R is then the pre-image of the diagonal under

$$p \times_k p \colon X \times_k X \longrightarrow wH(X) \times_k wH(X).$$

Using the exponential laws, we can recognise again that $p \times_k p$ is an identification. Because R is closed, the space $wH(X)$ must be a weak Hausdorff space. The equivalence relation is compatible with continuous maps because pre-images of closed equivalence relations are, again, closed. Overall, this results in the functor sH and the natural transformation $p \colon X \to wH(X)$. It remains for us to show that every continuous map f from X into a weak Hausdorff space Y factors through $wH(X)$. This is clear because the smallest closed equivalence relation on Y is the diagonal and thus is $Y \cong wH(Y)$. So $wH(f)$ provides the desired factorisation. The second statement follows again for general reasons. If T is a weak Hausdorff space and $f \colon wH(X') \to T$ is a map, for which $wH(q)f$ is continuous, then $qpf \colon X \to T$ is also continuous. Because p and q are identifications, the map f must be continuous, and the claim follows. □

Transformation Groups 5

In this chapter, we examine the symmetries of topological spaces. In this way, interesting new topological objects arise that deserve separate consideration. However, it is also possible to initially skim this chapter and later (for example, for Chap. 8) return to it. For a closer examination of the theory of transformation groups, we recommend [tD87].

5.1 Basic Concepts of Equivariant Topology

Definition 5.1.1

A *topological group* is a group G together with a topology on G, so that the maps

$$m: G \times G \longrightarrow G, \quad (g, h) \longmapsto gh$$

$$i: G \longrightarrow G, \quad g \longmapsto g^{-1}$$

are continuous.

Examples 5.1.2
Every group is a topological group with respect to the discrete topology. In this way, groups are often topologised. However, every group is also a topological group with respect to the clump topology. This is not so important. Typical examples of topological groups are given by matrix groups. We can consider the group $GL(2, \mathbb{R})$ of invertible $(2, 2)$–matrices as an open subspace of $\mathbb{R}^4$:

$$GL(2, \mathbb{R}) = \left\{ \begin{pmatrix} a & b \\ c & d \end{pmatrix} \mid ad - bc \neq 0 \right\}.$$

© The Author(s), under exclusive license to Springer-Verlag GmbH, DE, part of Springer Nature 2025
G. Laures, M. Szymik, *A Basic Course in Topology*, Compact Textbooks in Mathematics, https://doi.org/10.1007/978-3-662-70602-2_5

Multiplication and inversion are given by the formulas

$$\begin{pmatrix} a & b \\ c & d \end{pmatrix} \begin{pmatrix} a' & b' \\ c' & d' \end{pmatrix} = \begin{pmatrix} aa' + bc' & ab' + bd' \\ ca' + dc' & cb' + dd' \end{pmatrix}$$

and

$$\begin{pmatrix} a & b \\ c & d \end{pmatrix}^{-1} = \frac{1}{ad - bc} \begin{pmatrix} d & -b \\ -c & a \end{pmatrix},$$

so they are obviously continuous. The matrix groups $GL(n, \mathbb{R})$ and $GL(n, \mathbb{C})$ of higher rank are also topological groups, as well as their closed subgroups $SL(n, \mathbb{R})$ and $SL(n, \mathbb{C})$ of those matrices whose determinant is 1. The groups $O(n)$ of orthogonal matrices and $U(n)$ of unitary matrices as well as $SO(n)$ and $SU(n)$ are even compact. The group $U(1)$ consists of the $(1, 1)$–matrices of complex numbers of unit length; it is thus homeomorphic to the circle S^1.

▶ **Remark 5.1.3** Groups that are simultaneously manifolds and whose multiplication map is smooth are called *Lie groups*. They are also topological groups because the smoothness of the inversion follows from the implicit function theorem.

Topological groups always appear where continuous symmetries of topological spaces are examined. To do this, we consider group actions on spaces:

Definition 5.1.4

Let G be a topological group and X be a topological space. Then a *continuous action* of G on X is a continuous map

$$m \colon G \times X \to X,$$

written $(g, x) \mapsto gx$, that satisfies the rules $1x = x$, where 1 is the neutral element of G, and $g(hx) = (gh)x$. The pair (X, m) is then called a *G–space*. More precisely, this defines a left G–action and a left G–space. Analogously, we can also define right G–actions and right G–spaces with continuous maps

$$m \colon X \times G \longrightarrow X$$

that need to satisfy $x1 = x$ and $(xg)h = x(gh)$.

▶ **Remark 5.1.5** Using the formula

$$xg = g^{-1}x,$$

every left space can be transformed into a right space and vice versa.

5.1 Basic Concepts of Equivariant Topology

Examples 5.1.6
The topological group $GL(n, \mathbb{R})$ acts on $\mathbb{R}^n$ by multiplication. The same applies to the other matrix groups. A simple example of a G–space is G itself, considered as a G–space by means of multiplication. The group G acts on itself also by *conjugation*

$$(g, x) \mapsto gxg^{-1}.$$

If G is abelian, then this action is trivial in the following sense:

Definition 5.1.7

A G–space X is called *trivial* if $gx = x$ always holds.

If X is a G–space then every group element g provides a homeomorphism

$$X \to X, \; x \mapsto gx$$

of X, and the calculation rules from above say that the map adjoint to the action is a group homomorphism from G into the homeomorphism group. If X is locally compact, conversely, every continuous group homomorphism from G into the homeomorphism group defines an action. In this case, it is, therefore, irrelevant which standpoint we take. The question arises here whether we can choose for G the homeomorphism group itself and, as a map, the identity. The following theorem tells us if this is possible and thus provides further important examples for actions.

Theorem 5.1.8
Let X be a compact Hausdorff space. Then the group $\mathrm{Aut}(X)$ of homeomorphisms of X is a topological group for the compact-open topology. It acts continuously on X.

Proof. Since X is locally compact, the composition

$$\mathrm{Hom}(X, X) \times \mathrm{Hom}(X, X) \longrightarrow \mathrm{Hom}(X, X)$$

is continuous. By restriction, it follows that the group multiplication is continuous. Similarly, the continuity of the action follows from the continuity of the evaluation for locally compact spaces. It remains to show that the inversion

$$i : \mathrm{Aut}(X) \longrightarrow \mathrm{Aut}(X), \; f \mapsto f^{-1}$$

is continuous. If $K \subseteq X$ is compact and $U \subseteq X$ is open, then f^{-1} maps the set K to U if and only if $X \setminus U$ is mapped by f to $X \setminus K$. So we have

$$i^{-1}M(K, U) = M(X \setminus U, X \setminus K)$$

Fig. 5.1 Orbits of the orthogonal group O(2) in the plane $\mathbb{R}^2$

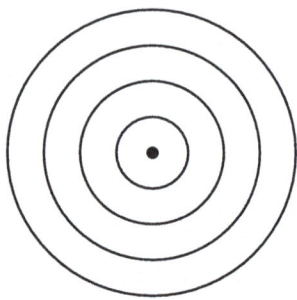

and, by assumption on X, the space $X \setminus U$ is compact and $X \setminus K$ is open. This shows the continuity of i. □

Definition 5.1.9

Let X and Y be two G–spaces. Then a continuous map $\varphi \colon X \to Y$ is a *continuous G–map* or *G–equivariant* if $\varphi(gx) = g\varphi(x)$ always holds.

The G–spaces together with the continuous G–maps form a category. In the case of the trivial group, this is the category of topological spaces.

Definition 5.1.10

Let X be a G–space and x a point in it. Then

$$G \longrightarrow X, \ g \longmapsto gx$$

is a continuous G–map, as it is the composite of the map from G to $G \times X$ that, on the first factor, is the identity and on the second is constant x with the action. Its image in X is the *orbit* Gx of x (Fig. 5.1). The pre-image of x in G is a subgroup, the *stabiliser* G_x of x. An action is called *free* if all stabilisers are trivial. In this case, the map is injective, thus providing a bijection from G to the orbit through x.

If X is a G–space, then the set

$$X/G = \{Gx \mid x \in X\}$$

of orbits becomes a quotient space of X so that the map $X \to X/G$ that assigns to each point its orbit is an identification. If $U \subseteq X$ is open, then

$$p^{-1}(p(U)) = \bigcup_{g \in G} gU$$

5.1 Basic Concepts of Equivariant Topology

is open, thus $p(U)$ is open. This shows:

▶ **Remark 5.1.11** The quotient map $X \to X/G$ is always open.

The quotient map is thus an identification with special properties.

Supplement

Representations A *representation* of a topological group G is a left action of G on a vector space V for which all maps

$$L_g : V \longrightarrow V, \ v \mapsto gv$$

are linear. If V is the vector space $\mathbb{C}^n$, then a representation corresponds to a continuous group homomorphism

$$G \longrightarrow \mathrm{GL}(n, \mathbb{C}).$$

If this map is injective, it is referred to as a *faithful representation*, and each element is then uniquely represented by the corresponding matrix. The theory of representations is particularly well understood for finite groups and compact Lie groups. For example, a theorem by Peter and Weyl states that every compact Lie group has a faithful representation and thus is a closed subgroup of $\mathrm{GL}(n, \mathbb{C})$, see for example [BtD95, Thm. 4.1].

Exercises

Exercise 82 Separately Grouped
Let G be a group with a topology that has the separation property (T1). Show that G is a topological group if and only if the map

$$G \times G \longrightarrow G, \ (g, h) \mapsto gh^{-1}$$

is continuous.

Exercise 83 Multiplying Accumulation Points
If H is a subgroup of a topological group G, then H and $\overline{H}$ together with the subspace topology and multiplication in G are topological groups.

Exercise 84 Partially Clumped and Radial
Describe the orbit space $\mathbb{C}/\mathrm{GL}(1, \mathbb{C})$. Also show that $\mathbb{C}/U(1)$ is homeomorphic to the interval $[\,0, \infty\,[$.

Exercise 85 Loop Groups
Let G be a topological group and X a topological space. Then the mapping space $\mathrm{Hom}(X, G)$ is a topological group with respect to pointwise multiplication: $(f \cdot f')(x) = f(x) \cdot f'(x)$. For $X = S^1$, this is the *loop group* of G.

Exercise 86 As it Should Be
Every continuous G–map $\varphi \colon X \to Y$ induces a continuous map $\varphi/G \colon X/G \to Y/G$ between the orbit spaces.

5.2 Homogeneous Spaces

Let H be a subgroup of a topological group G. Then H acts by right multiplication on G, and the orbit space G/H is the space of the orbits, which have the form

$$gH = \{gh | h \in H\}.$$

The orbits are always homeomorphic to each other, unlike for general actions. They are also called *right cosets*. Note that the notation G/H is also used for the collapsing of H in G to a point. This is rarely meant in the context of topological groups. The difference should be clear from the example $\mathbb{R}/\mathbb{Z}$.

Theorem 5.2.1
The space G/H is a Hausdorff space if and only if H is closed in G.

Proof. If G/H is a Hausdorff space then the singleton subset $\{1H\}$ in G/H is closed, so its pre-image H under the continuous map $p \colon G \to G/H$ is so too. Now let H be closed. Then the pre-image of H under the continuous map

$$G \times G \longrightarrow G, \ (g, g') \longmapsto g^{-1}g'$$

is closed. This is the set of pairs (g, g') with $gH = g'H$, so the equivalence relation that belongs to the surjection p. Since p is also open, the closedness of the equivalence relation implies that the quotient is a Hausdorff space. □

Theorem 5.2.2
The quotient map $p \colon G \to G/H$ is proper if and only if H is compact.

Proof. If p is proper, then the pre-image H of $\{1H\}$ is compact. Let now conversely H be compact. The fibres of p are the cosets of H, so compact. It remains to show that p is closed. Since p is an identification, it suffices to show that for every

5.2 Homogeneous Spaces

closed A in G, the set $p^{-1}(p(A))$ is closed in G. Now is $p^{-1}(p(A)) = AH$. Let g be in the complement of it. For all h in H, the subset $G \setminus Ah$ is open and contains g. So, there are open neighbourhoods $U(h)$ of g and $V(h)$ of 1 with $U(h)V(h) \subseteq G \setminus Ah$. Hence, we have $U(h)V(h) \cap Ah = \emptyset$ and this implies $U(h) \cap AhV(h)^{-1} = \emptyset$. Since H is compact there is a finite subset $H_0 \subseteq H$ with

$$H \subseteq \bigcup_{h \in H_0} hV(h)^{-1}.$$

Let

$$U = \bigcap_{h \in H_0} U(h).$$

Then U is an open neighbourhood of g. If it met AH, then there would be a u in U of the form $u = ah$ with an a from A and a h from H. Now, however, the h lies in $h_0 V(h_0)^{-1}$ for a h_0 in H_0 and so $ah \in A h_0 V(h_0)^{-1}$. This set is disjoint to $U(h_0) \supseteq U$. □

By left multiplication, the set G/H is a G–space. The continuity can be seen from the commutative diagram

$$\begin{array}{ccc} G \times G & \xrightarrow{m} & G \\ {\scriptstyle \mathrm{id}_G \times p} \downarrow & & \downarrow {\scriptstyle p} \\ G \times G/H & \longrightarrow & G/H. \end{array}$$

Since the map p is not only an identification but also open, the map $\mathrm{id}_G \times p$ is open, thus an identification, and pm is continuous.

Definition 5.2.3

A G–space X that is G–homeomorphic to one of the G–spaces G/H is called a *homogeneous G–space*.

A G–set X is *transitive* if it consists of a unique orbit. On homogeneous G–spaces, the group G acts transitively. If N is a *normal subgroup* in G, that is,

$$gN = Ng$$

for all g, then G/N is not only a space but also a group. The group multiplication in G/N is then defined by

$$(gN)(hN) = (gh)N$$

With a similar diagram argument, we can show that G/N is also a topological group. Suppose X is a G–space and let x be a point in it. Then the map $G \to X$, $g \mapsto gx$ induces a continuous G–bijection

$$G/G_x \to Gx.$$

In many examples, this map is a G–homeomorphism.

Examples 5.2.4
A beautiful class of examples is provided by linear algebra. The unitary group $U(n)$ acts on $\mathbb{C}^n$ by multiplying matrices with vectors. This action is not transitive, but the restriction to the sphere S^{2n-1} in $\mathbb{C}^n$ is because every vector of unit length can be extended to an orthonormal basis. The stabiliser of such a vector consists of the unitary transformations of the orthogonal complement so that it can be identified with $U(n-1)$. We get

$$U(n)/U(n-1) \cong S^{2n-1},$$

because the above map is a continuous bijection from a compact to a Hausdorff space. With real matrices, we get

$$O(n)/O(n-1) \cong S^{n-1}.$$

More generally, we consider the action of $U(n)$ on the k–fold product

$$(\mathbb{C}^n)^k = \mathbb{C}^n \times \cdots \times \mathbb{C}^n$$

through component-wise action:

$$A(v_1, \ldots, v_k) = (Av_1, \ldots, Av_k).$$

Now, let $V_k(\mathbb{C}^n) \subseteq \mathbb{C}^n \times \cdots \times \mathbb{C}^n$ be the subset of orthonormal systems, which in this context are also called *k–legs*. Then $U(n)$ acts transitively on $V_k(\mathbb{C}^n)$, and the stabiliser of a k–leg consists of the unitary transformations of the orthogonal complement; it can, therefore, be identified with $U(n-k)$. It follows

$$U(n)/U(n-k) \cong V_k(\mathbb{C}^n).$$

The spaces $V_k(\mathbb{C}^n)$ are called *Stiefel manifolds*. The *Grassmann manifold* $G_k(\mathbb{C}^n)$ is the set of k–dimensional subvector spaces of $\mathbb{C}^n$. Since every such subvector space has an orthonormal basis, there is a surjective map

$$V_k(\mathbb{C}^n) \longrightarrow G_k(\mathbb{C}^n)$$

5.2 Homogeneous Spaces

that assigns to each k–leg the subvector space generated by it. This map makes the Grassmann manifold a quotient of the Stiefel manifold. A bit of linear algebra shows that the subgroup $U(k)$ of $U(n)$ acts freely on $V_k(\mathbb{C}^n)$. The above map induces a homeomorphism

$$V_k(\mathbb{C}^n)/U(k) \cong G_k(\mathbb{C}^n).$$

Together with the representation of the Stiefel manifold as a homogeneous space, we thus obtain a homeomorphism

$$U(n)/(U(k) \times U(n-k)) \cong G_k(\mathbb{C}^n).$$

The case $k = 1$ is particularly prominent. The space $G_1(\mathbb{C}^n)$ is the space of lines in $\mathbb{C}^n$, also known as the *complex projective space* $\mathbb{C}P^{n-1}$ of complex dimension $n - 1$. As above, there is a homeomorphism

$$S^{2n-1}/U(1) \cong \mathbb{C}P^{n-1}.$$

The image of a vector $(z_1, \ldots, z_n)$ is denoted by $[z_1, \ldots, z_n]$. These are the *homogeneous coordinates* of the image point. Two points $(z_1, \ldots, z_n)$ and $(z'_1, \ldots, z'_n)$ describe the same point in $\mathbb{C}P^n$ if there is an a in $U(1)$ so that $z'_j = a z_j$ holds for all j. The projective space $\mathbb{C}P^{n-1}$ can also be realised as a quotient of $\mathbb{C}^n \setminus 0$ with respect to the action of $GL(1, \mathbb{C})$ by scalar multiplication. This point of view has the disadvantage that the projective space is not immediately recognised as compact; however, it is immediately recognised that the scalar product plays no essential role in the construction. In the real case, we obtain the real projective space $\mathbb{R}P^{n-1}$, which was already introduced in Example 2.4.2:

$$\mathbb{R}P^{n-1} = (\mathbb{R}^n \setminus 0)/GL(1, \mathbb{R}) \cong S^{n-1}/O(1).$$

Definition 5.2.5

Let X be a G–space. For each g in G, we consider the set

$$X^g = \{x \in X \mid gx = x\}$$

of points that are left fixed by g. If H is a subgroup of G, the *H–fixed point space* is defined by

$$X^H = \bigcap_{h \in H} X^h.$$

Example 5.2.6

The group of invertible matrices $GL(3, \mathbb{R})$ contains the subgroup of matrices whose last column and row only have the value 1 on the diagonal and otherwise vanish. This group is isomorphic to $GL(2, \mathbb{R})$, and the fixed point space under the action on $\mathbb{R}^3$ is the third axis.

The following theorem shows the close relationship between quotient groups and fixed point spaces.

> **Theorem 5.2.7**
> Let G be a compact group and H a closed subgroup. Then the space $\mathrm{Hom}_G(G/H, X)$ of continuous G–maps from G/H to X is homeomorphic to the H–fixed point space X^H.

Proof. It is easy to verify that

$$\alpha \colon \mathrm{Hom}_G(G/H, X) \longrightarrow X^H, \quad f \longmapsto f(1H)$$

and

$$\beta \colon X^H \longrightarrow \mathrm{Hom}_G(G/H, X), \quad x \longmapsto (gH \mapsto gx)$$

are well-defined, mutually inverse maps. It remains to show that these are continuous. The map α is a restriction of an evaluation map. Since G/H, as a compact Hausdorff space, is also locally compact, this is continuous. The map adjoint to β is the map $X^H \times G/H \to X$ that maps (x, gH) to gx, so it is continuous as well. Then β is also continuous. $\square$

The homeomorphisms α and β given in the proof are, by the way, *natural* in the following sense: if $\varphi \colon X \to Y$ is a continuous G–map, then the diagram

$$\begin{array}{ccc} \mathrm{Hom}_G(G/H, X) & \xrightarrow{\alpha_X} & X^H \\ \varphi_* \downarrow & & \downarrow \varphi^H \\ \mathrm{Hom}_G(G/H, Y) & \xrightarrow{\alpha_Y} & Y^H \end{array}$$

is commutative. The same applies to β.

Supplements

The Weyl Group For a subgroup H of G we define the *normaliser* NH by

$$NH = \{n \in G \mid nH = Hn\}.$$

The normaliser is the maximal subgroup of G that contains H as a normal subgroup. The *Weyl group* is the quotient group

$$WH = NH/H.$$

5.2 Homogeneous Spaces

The Weyl group acts on G/H from the right by $(gH)[n] = gnH$. The associated translations

$$R_{[n]} \colon G/H \to G/H, \ gH \mapsto gnH$$

are G–equivariant. Conversely, any G–self-map of G/H is of this form for a $[n] \in G/H$. Any such n necessarily fulfils $n^{-1}Hn \subseteq H$ because then $hnH = nH$ for every $h \in H$. If the self-map is injective then $n^{-1}H = hn^{-1}H$ also holds for all $h \in H$, and thus we have $nHn^{-1} \subseteq H$. Overall, then $n \in NH$, and the Weyl group corresponds to the G–automorphism group of G/H.

Symmetric Products The *symmetric group* Σ_n is the group of bijections of the set $\{1, 2, \ldots, n\}$. It acts on the n–fold product X^n of any topological space X by swapping the factors

$$((x_1, x_2, \ldots, x_n), \sigma) \mapsto (x_{\sigma(1)}, x_{\sigma(2)}, \ldots, x_{\sigma(n)}).$$

The orbit space is called the n–fold *symmetric product* $\mathrm{SP}^n(X)$. For example, the space $\mathrm{SP}^n(\mathbb{CP}^1)$ is homeomorphic to $\mathbb{CP}^n$ by

$$[[a_0, b_0], \ldots, [a_n, b_n]] \mapsto [c_0, \ldots, c_n],$$

where the numbers c_i are determined by the equation

$$\prod_{i=0}^{n}(a_i x + b_i y) = c_0 x^n + c_1 x^{n-1} y + \cdots + c_n y^n.$$

Beware: the symmetric product $\mathrm{SP}^n(\mathbb{RP}^1)$ is not homeomorphic to $\mathbb{RP}^n$; it maps to the circle so that the fibres are $(n-1)$–simplices (see Sect. 11.2).

Exercises

Exercise 87 Homogeneous Among Each Other
Let G be a topological group and let H and K be subgroups of G. Then every G–map $G/H \to G/K$ is continuous.

Exercise 88 Irrational
The quotient group $\mathbb{R}/\mathbb{Q}$ carries the clump topology.

Exercise 89 Homogeneous and Connected
Let H be a connected subgroup of G. Show that G is connected if and only if G/H is.

Exercise 90 A Thick Point
Let G be a topological group. Then the closure N of $\{1\}$ is a normal subgroup. If X is a Hausdorff G–space, then N lies in every stabiliser of the action.

Exercise 91 Closed Subgroups
Show that the Stiefel and Grassmann manifolds are Hausdorff spaces.

Exercise 92 Flags
Let $F(\mathbb{C}^n) \subseteq G_1(\mathbb{C}^n) \times \cdots \times G_{n-1}(\mathbb{C}^n)$ be the set of $(n-1)$–tuples $(U_1, \ldots, U_{n-1})$ with $U_j \subseteq U_{j+1}$ for all j. The group $U(n)$ acts component-wise on $F(\mathbb{C}^n)$. Is this action transitive? What do the stabilisers look like?

Exercise 93 A Branch
The subgroup $\{z \in GL(1, \mathbb{C}) \mid z^n = 1\}$ of the n–th roots of unity acts on $\mathbb{C}$ through scalar multiplication. Show that the orbit space is homeomorphic to $\mathbb{C}$.

Exercise 94 Fix
If X is a Hausdorff G–space, then the fixed point sets X^H are closed for all subgroups $H \subseteq G$.

5.3 Proper Actions

When transitioning from a G–space X to the orbit space X/G, two points are identified if they are in the same orbit. This equivalence relation R is a subspace of $X \times X$. It often proves useful not only to remember *what* is being identified but also *how* it is being identified. In the present case, this is done by considering the continuous map

$$\theta \colon G \times X \longrightarrow X \times X, \ (g, x) \longmapsto (x, gx)$$

whose image is R. In particular, the equivalence relation R is determined by θ, but not vice versa.

Definition 5.3.1

An action is called *proper* if the map θ is proper.

We postpone the examples to learn more about proper actions first.

Theorem 5.3.2
If X is a proper G–space, then X/G is a Hausdorff space.

Proof. It suffices to show that the relation R is closed in $X \times X$. This is the case because R is the image of the proper (thus closed) map θ. □

5.3 Proper Actions

Theorem 5.3.3
Let X be a proper G–space and x be a point from X. Then the stabiliser G_x is compact and the orbit Gx is closed. The map

$$G \longrightarrow X, \; g \longmapsto gx$$

is proper and induces a G–homeomorphism $G/G_x = Gx$.

Proof. The map arises from θ by base change along the injection

$$X \longrightarrow X \times X, \; x' \longmapsto (x, x')$$

and therefore is proper. The stabiliser is the pre-image of a point, hence compact. The orbit is the image of a closed set, so closed. Finally, it follows that the induced map $G/G_x \to Gx$ is also closed. It is continuous and bijective anyway. □

Now, we present some criteria that enable us to prove many group actions to be proper.

Theorem 5.3.4
Let G be a compact Hausdorff group and X a Hausdorff G–space. Then G acts properly on X.

Proof. Since G is compact, the projection

$$\mathrm{pr}_X : G \times X \longrightarrow X, \; (g, x) \longmapsto x$$

is proper. Then the action

$$m : G \times X \longrightarrow X, \; (g, x) \longmapsto gx$$

is proper because it is the composite of the projection pr_X with the homeomorphism $(g, x) \mapsto (g, gx)$ from $G \times X$. Hence, the product of the projection and the action is proper. The map θ is the composite of this product with the diagonal map of $G \times X$:

$$G \times X \xrightarrow{\Delta} G \times X \times G \times X \xrightarrow{\mathrm{pr}_X \times m} X \times X \; .$$

It is, therefore, sufficient to show that this diagonal is proper. Since it is injective, this is equivalent to the statement that it is closed. It is always an embedding because it has a one-sided inverse. Therefore, it is closed if and only if its image is closed. This, however, is equivalent to the statement that $G \times X$ is a Hausdorff space. □

Let X be a free G–space. Then θ induces a continuous bijection

$$\theta': G \times X \longrightarrow R.$$

The inverse map $R \to G \times X$ has the form

$$(x, x') \longmapsto (\varphi(x, x'), x)$$

for a map $\varphi \colon R \to G$. It assigns to each pair (x, x') from R the (uniquely determined) group element g that fulfils $x' = gx$.

Theorem 5.3.5
Let X be a free G–space. Then the following are equivalent:

(a) *The group G acts properly on X.*
(b) *The relation R is closed, and θ' is a homeomorphism.*
(c) *The relation R is closed, and φ is continuous.*

Proof. The map θ' is a homeomorphism if the inverse map is continuous. Therefore, we see that (b) and (c) are equivalent. On the other hand, the map θ' is a homeomorphism if and only if it is closed. Since R is closed, this is equivalent to θ being closed, thus proper. Consequently, we find that (a) and (b) are also equivalent. □

Theorem 5.3.6
The G–space G/H is proper if and only if H is compact.

Proof. Since H is the stabiliser of $1H$, compactness of H is certainly necessary for being proper. For the converse, let us first clarify the case $H = 1$. For the action of G on itself, the relation $R = G \times G$ is closed. Furthermore, the map φ is given by $\varphi(g, h) = hg^{-1}$, thus continuous. The previous theorem now implies that G acts

5.3 Proper Actions

properly on itself. The general case can be reduced to the special case using the diagram

$$
\begin{array}{ccc}
G \times G & \xrightarrow{\theta_G} & G \times G \\
{\scriptstyle \mathrm{id}_G \times p} \downarrow & & \downarrow {\scriptstyle p \times p} \\
G \times G/H & \xrightarrow{\theta_{G/H}} & G/H \times G/H.
\end{array}
$$

Since H is compact, the map p is proper and hence so are $\mathrm{id}_G \times p$ and $p \times p$. In the special case, it was shown that θ_G is proper. Then the map $\theta_{G/H}$ also is proper because of the surjectivity of $\mathrm{id}_G \times p$. □

Example 5.3.7
Finally, we mention another example that is an essential piece of general mathematical knowledge. Let

$$H = \{z \in \mathbb{C} \mid \mathrm{Im}(z) > 0\}$$

be the upper half-plane. The group $\mathrm{SL}(2, \mathbb{R})$ acts on H by

$$\begin{pmatrix} a & b \\ c & d \end{pmatrix} z = \frac{az+b}{cz+d}.$$

This action can, by the way, be understood as a restriction of the canonical action of $\mathrm{GL}(2, \mathbb{C})$ on $\mathbb{CP}^1$. For $y > 0$ we have

$$\begin{pmatrix} \sqrt{y} & x/\sqrt{y} \\ 0 & 1/\sqrt{y} \end{pmatrix} \mathrm{i} = \frac{\sqrt{y}\,\mathrm{i} + x/\sqrt{y}}{1/\sqrt{y}} = x + \mathrm{i}y.$$

Consequently, every point lies in the orbit of i; the group $\mathrm{SL}(2, \mathbb{R})$ acts transitively on H. The stabiliser of i is $\mathrm{SO}(2)$. (Verify this!) We obtain a continuous bijection between $\mathrm{SL}(2, \mathbb{R})/\mathrm{SO}(2)$ and H. The matrix with the root entries shows that the inverse map is continuous. Thus there is an $\mathrm{SL}(2, \mathbb{R})$–homeomorphism

$$\mathrm{SL}(2, \mathbb{R})/\mathrm{SO}(2) \cong H.$$

This also shows that the action of $\mathrm{SL}(2, \mathbb{R})$ on H is proper. The group $\mathrm{SL}(2, \mathbb{R})$ has many discrete subgroups, such as $\mathrm{SL}(2, \mathbb{Z})$. These act (by restriction) also properly on the upper half-plane and provide very interesting quotients.

Exercises

Exercise 95 This Is Typical Again
The group $\mathbb{Z}$ acts on the space $\mathbb{R}$ by translation: $(g, x) \mapsto g + x$. This action is free and proper.

Exercise 96 A Double Point
The group $\mathrm{GL}(1, \mathbb{R}) = \mathbb{R} \setminus 0$ acts on the punctured plane $\mathbb{R}^2 \setminus 0$ via $t(x, y) = (tx, t^{-1}y)$. The orbits are closed. The action is free but not proper. The map $(x, y) \mapsto xy$ a defines a local homeomorphism from the orbit space to $\mathbb{R}$. The orbit space is not a Hausdorff space.

Exercise 97 Properly Inherited
Let H be a closed subgrup of G. Let X be a space on which G acts properly. Then also H acts properly on X by restricting the action of G.

Exercise 98 Properly Descended
Suppose G acts properly on X and suppose $f : X \to Y$ is a surjective proper G–map. Then G also acts properly on Y.

Paths and Loops 6

Topological problems are, in general, too complicated to be solved directly. A rather crude method to make them more manageable is to discretise them by replacing the spaces with the sets of their path components. Finer discretisation methods are obtained by first replacing the spaces with auxiliary spaces and then moving on to the path components. The resulting sets often have an algebraic structure, making their description easier. We will hint at this at the end of this chapter using the example of the circle and then exploit it further in the following chapters.

6.1 Path Spaces and Loop Spaces

Definition 6.1.1

Let X be a topological space. A *path* in X is a continuous map

$$\gamma: I \longrightarrow X$$

from the compact interval $I = [0, 1]$ to X. We also speak of a path from $\gamma(0)$ to $\gamma(1)$. We denote by PX the *path space* $\mathrm{Hom}(I, X)$, i.e., the space of all paths in X. It is, of course, equipped with the compact-open topology.

The path space is to be considered as an auxiliary space in the study of X. In addition to its topology, it carries another structure given by the following constructions. The evaluations at the two endpoints are continuous maps

$$\mathrm{ev}_0 \colon PX \longrightarrow X, \ \gamma \longmapsto \gamma(0),$$
$$\mathrm{ev}_1 \colon PX \longrightarrow X, \ \gamma \longmapsto \gamma(1).$$

© The Author(s), under exclusive license to Springer-Verlag GmbH, DE, part of Springer Nature 2025
G. Laures, M. Szymik, *A Basic Course in Topology*, Compact Textbooks in Mathematics, https://doi.org/10.1007/978-3-662-70602-2_6

Fig. 6.1 Composition of paths

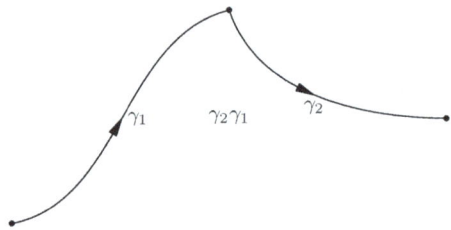

For each point, there is a constant path, and this provides the continuous map

$$e\colon X \longrightarrow PX,\ x \longmapsto (t \mapsto x).$$

For each path γ there is an inverse path γ^-, defined as the composition of the inversion $t \mapsto 1-t$ with γ. This path has the same image but is traversed in the opposite direction. In particular, the starting point and endpoint are swapped. This provides a continuous map

$$i\colon PX \longrightarrow PX,\ \gamma \longmapsto \gamma^-.$$

Finally, paths can be composed, but only if the endpoint of the first path is the starting point of the second path. If γ_1 and γ_2 are paths in X with $t(\gamma_1) = s(\gamma_2)$, a path is defined by

$$\gamma_2\gamma_1 = \begin{cases} \gamma_1(2t) & t \in [0, 1/2] \\ \gamma_2(2t-1) & t \in [1/2, 1], \end{cases}$$

where first γ_1 and then γ_2 are traversed, but each at double speed, so that the entire path is completed during the interval $[0,1]$ (see Fig. 6.1).

This composition provides a map

$$m\colon PX \times_X PX \longrightarrow PX, (\gamma_1, \gamma_2) \longmapsto \gamma_2\gamma_1$$

that is defined on the pullback of the maps ev_1 and ev_0.

▶ **Remark 6.1.2** The map m is a homeomorphism. This can be deduced from the commutative diagram

$$\begin{array}{ccc}
\mathrm{Hom}([0,\tfrac{1}{2}], X) \times_{\mathrm{Hom}(\{\tfrac{1}{2}\}, X)} \mathrm{Hom}([\tfrac{1}{2}, 1], X) & \longrightarrow & \mathrm{Hom}([0,\tfrac{1}{2}] +_{\{\tfrac{1}{2}\}} [\tfrac{1}{2}, 1], X) \\
\subseteq \downarrow & & \downarrow \\
\mathrm{Hom}([0,\tfrac{1}{2}], X) \times \mathrm{Hom}([\tfrac{1}{2}, 1], X) & \longrightarrow & \mathrm{Hom}([0,\tfrac{1}{2}] + [\tfrac{1}{2}, 1], X).
\end{array}$$

6.1 Path Spaces and Loop Spaces

The lower map is the bijection from the universal property of the sum. It is a homeomorphism because it is continuous, as the associated adjoint is, and its inverse is continuous because their projections are the compositions with the respective inclusions. The left map is the inclusion of a subspace. To show that the upper bijection, given by the universal property of the pushout, is also a homeomorphism, it is sufficient to show that the right map is an embedding. It is induced by the identification

$$[0, 1/2] + [1/2, 1] \longrightarrow [0, 1/2] +_{\{1/2\}} [1/2, 1],$$

thus, it is injective and continuous. It is an embedding because, according to the exponential law, it has the universal property that characterises embeddings. The map m finally differs from the upper one only by homeomorphisms; therefore, it is itself one.

Some variants of path spaces arise from the choice of starting or endpoints. We would only reluctantly have the agony of choice, but sometimes, it is so helpful that we accept the evil as necessary.

Definition 6.1.3

If x_0 and x_1 are points in X, then we denote by

$$PX(x_0, x_1)$$

the subspace of PX of the paths from x_0 to x_1.

The space $PX(x_0, x_1)$ is the fibre of the continuous map

$$(\mathrm{ev}_0, \mathrm{ev}_1) \colon PX \longrightarrow X \times X, \ \gamma \longmapsto (\gamma(0), \gamma(1))$$

above the point (x_0, x_1). (As a set, the space PX is the disjoint union of the spaces $PX(x_0, x_1)$; but it is not their sum in the sense of topology.) The case $x_0 = x_1$ is distinguished and deserves its own notation.

Definition 6.1.4

If x is a point of X, then

$$\Omega(X, x) = PX(x, x)$$

is called the *loop space* of X at x. A *loop* in X at x is thus a continuous path that begins and ends in x. The *free loop space*

$$LX = \text{Hom}(S^1, X)$$

is the space of all continuous maps of the circle S^1 to X. (As a set, the space LX is the disjoint union of the spaces $\Omega(X, x)$; but it is not their sum in the sense of topology.)

This might also be a good time to note that the notation PX is used in many books for a different space. If x is a point of X, then we can consider the subspace of PX that consists of the paths that start in x. This space is sometimes denoted by PX as well, although it also depends on x, and it would be more accurate to denote it by $P(X, x)$. This is what we will do here. Then $\Omega(X, x)$ is a subspace of $P(X, x)$, namely the fibre of the map to X that maps each path to its endpoint.

Supplement

Path Category and Moore Paths The space PX is closely related with a category $\mathcal{P}X$ of paths in X. However, in the definition of paths in the path category, different lengths for the intervals are allowed to make the concatenation of paths strictly associative. The objects of $\mathcal{P}X$ are points in X. Morphisms from x_0 to x_1 are pairs (l, γ), where $0 \leqslant l < \infty$ is a real number and

$$\gamma : [0, l] \longrightarrow X$$

is a continuous map with $\gamma(0) = x_0$ and $\gamma(l) = x_1$. The composition of two such paths is then

$$(l', \gamma')(l, \gamma) = (l + l', \gamma' \cdot \gamma),$$

where $\gamma' \cdot \gamma : [0, l + l'] \to X$ is given by

$$t \longmapsto \begin{cases} \gamma(t) & \text{for } t \in [0, l] \\ \gamma'(t - l) & \text{for } t \in [l, l + l']. \end{cases}$$

This composition is associative. The identity id_x for the object x is the path defined on the interval of length 0 with value x.

Exercises

Exercise 99 Retracted Paths
Let $p\colon I \to S^1$, $t \mapsto \exp(2\pi i t)$ be the identification of the two endpoints. Show that the diagram

$$\begin{array}{ccc} LX & \xrightarrow{p^*} & PX \\ {\scriptstyle ev_1}\downarrow & & \downarrow{\scriptstyle (ev_0, ev_1)} \\ X & \xrightarrow{\Delta} & X \times X \end{array}$$

fulfils the universal property of the pullback from Theorem 2.2.8. In particular, the fibre $ev_1^{-1}\{x\}$ of the continuous evaluation

$$ev_1\colon LX \longrightarrow X, \quad f \mapsto f(1)$$

over x is homeomorphic to $\Omega(X, x)$.

6.2 The Path Component Functor

In this section, we use path spaces to understand further properties of topological spaces.

Definition 6.2.1

Two points x_0 and x_1 of a topological space X are called *connectable* if there is a path in X from x_0 to x_1. This is the case if and only if the space $PX(x_0, x_1)$ is not empty. The space X is called *path-connected* if any two points in X can be connected by a path.

Not every connected space is path-connected, as can be seen from the example of the sine space from Sect. 3.1. The converse, however, is true:

Theorem 6.2.2
Every path-connected space is connected.

Proof. Suppose there is a continuous map $f\colon X \to S^0$ with $f(x_0) = 1$ and $f(x_1) = -1$ for certain points $x_0, x_1 \in X$. Then we find a path γ that connects x_0 with x_1. The composition $f\gamma$ is a continuous map with a connected domain whose image is not connected. □

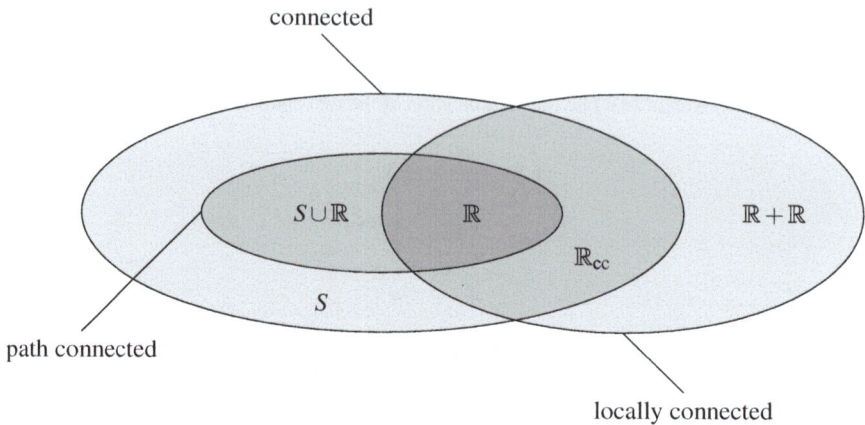

Fig. 6.2 Implications between different connectivity concepts

Figure 6.2 provides an overview of the implications between different connectivity concepts. Here, the space S is the sine space, and $\mathbb{R}_{cc}$ is the set of real numbers with the topology of countable complements.

From the geometric constructions of the last section, especially the maps e, i, and m, we immediately see the following.

▶ **Remark 6.2.3** Path-connectivity is an equivalence relation.

Definition 6.2.4

The equivalence class $[x]$ of x is called the *path component* of x. The set of path components is denoted by $\pi_0 X$.

If $f\colon X \to Y$ is a continuous map and γ is a path that connects x_0 with x_1, then $f\gamma$ is a path that connects $f(x_0)$ with $f(x_1)$. This defines a map

$$\pi_0 f \colon \pi_0 X \to \pi_0 Y$$

by $[x] \mapsto [f(x)]$. Assignments between categories with these properties occur very frequently. Therefore, the following terminology is introduced.

Definition 6.2.5

Let $\mathcal{C}$ and $\mathcal{D}$ be categories. A *functor* F from $\mathcal{C}$ to $\mathcal{D}$ is an assignment that assigns to each object $X \in \mathcal{C}$ an object FX in $\mathcal{D}$ as well as to each morphism

6.2 The Path Component Functor

$f \in \mathrm{Mor}_{\mathcal{C}}(X, Y)$ a morphism $Ff \in \mathrm{Mor}_{\mathcal{D}}(FX, FY)$, such that the properties

(FA1) $F\mathrm{id}_X = \mathrm{id}_{FX}$ for all objects X from $\mathcal{C}$
(FA2) $F(gf) = (Fg)(Ff)$ for all composable morphisms f, g from $\mathcal{C}$

hold.

Example 6.2.6
Suppose $\mathcal{C}$ and $\mathcal{D}$ each contain only one object, and every morphism is invertible. Then the sets of morphisms are groups, and a functor corresponds to a group homomorphism according to (FA2).

Example 6.2.7
The assignments

$$\pi_0 \colon \mathbf{Top} \longrightarrow \mathbf{Sets}$$
$$X \longmapsto \pi_0 X$$
$$f \longmapsto \pi_0 f.$$

define a functor, the *path component functor*. Even the construction of the path space

$$P \colon \mathbf{Top} \longrightarrow \mathbf{Top}$$

is functorial. If $f \colon X \to Y$ is a continuous map, then we have a continuous map

$$Pf \colon PX \to PY, \gamma \longmapsto f\gamma,$$

that satisfies the functor axioms $P(\mathrm{id}) = \mathrm{id}$ and $P(fg) = P(f)P(g)$.

While at first glance perhaps inconspicuous, the construction of the path component functor is actually one of the most important constructions in topology. It provides a transition from the continuous world of topological spaces to the discrete world of sets. From a certain perspective, every such transition crossed this bridge. This requires considering path components of suitable auxiliary spaces.

▶ **Remark 6.2.8** If $f \colon X \to Y$ is surjective, so is $\pi_0 f \colon \pi_0 X \to \pi_0 Y$.

This statement becomes false if 'surjective' is replaced by 'injective'. This can be seen, for example, for the inclusion of $\{-1, +1\}$ to S^1. The inclusion is injective, of course, but not the induced map on the path components. By the way, the commutative diagram

$$\begin{array}{ccc} \{-1, +1\} & \xrightarrow{\subseteq} & S^1 \\ \downarrow & & \downarrow {\scriptstyle z \mapsto z^2} \\ \{1\} & \xrightarrow{\subseteq} & S^1 \end{array}$$

is a pullback, as it fulfils the universal property of Theorem 2.2.8. In other words, the maps from the diagram induce a homeomorphism

$$\{-1, +1\} \longrightarrow \{1\} \times_{S^1} S^1.$$

Thus, it is not true that π_0 maps pullbacks of spaces to pullbacks of sets. As another example, we consider the two subsets

$$U = \mathbb{R}^2 \setminus \{(0, y) \mid y \geq 0\} \quad \text{and} \quad V = \mathbb{R}^2 \setminus \{(0, y) \mid y \leq 0\}$$

of the plane. Then $U \cap V = \mathbb{R}^2 \setminus \mathbb{R}$ and $U \cup V = \mathbb{R}^2 \setminus 0$. The commutative diagram

$$\begin{array}{ccc} U \cap V & \xrightarrow{\subseteq} & V \\ {\scriptstyle \subseteq} \downarrow & & \downarrow {\scriptstyle \subseteq} \\ U & \xrightarrow{\subseteq} & U \cup V \end{array}$$

is a pullback, but it does not remain a pullback when π_0 is applied to it. The problem lies in the incompatibility with inclusions because π_0 is compatible with products, as the following result shows.

> **Theorem 6.2.9**
> Let X and Y be topological spaces. Then the projections induce a bijection
> $$\pi_0(X \times Y) \xrightarrow{\cong} \pi_0(X) \times \pi_0(Y).$$

Proof. The map is given by

$$[(x, y)] \longmapsto ([x], [y])$$

so it is surjective. If $([x_0], [y_0]) = ([x_1], [y_1])$, then there is a path γ in X from x_0 to x_1 and a path γ' in Y from y_0 to y_1. Then is $(\gamma, \gamma') \colon I \to X \times Y$ a path from (x_0, y_0) to (x_1, y_1), thus $[x_0, y_0] = [x_1, y_1]$. This shows injectivity. □

Example 6.2.10
If G is a topological group, then $\pi_0 G$ is a group through multiplication

$$\pi_0(G) \times \pi_0(G) \cong \pi_0(G \times G) \longrightarrow \pi_0(G), \quad ([g], [h]) \longmapsto [gh].$$

6.2 The Path Component Functor

For a particularly important class of examples, if $G = \mathrm{Aut}(X)$ is the homeomorphism group of a compact Hausdorff space X, then $\pi_0 \mathrm{Aut}(X)$ is called the *mapping class group* of X. Understanding it is the first step in studying $\mathrm{Aut}(X)$ itself.

Theorem 6.2.11
Let X and Y be topological spaces. Then the inclusions induce a bijection
$$\pi_0(X) + \pi_0(Y) \xrightarrow{\cong} \pi_0(X+Y).$$

Proof. Surjectivity is again clear. Injectivity results directly from the fact that every path in $X + Y$ either runs in X or in Y. $\square$

We can also ask whether more general pushouts are mapped onto pushouts by π_0. There is always a map
$$\pi_0(X) +_{\pi_0(A)} \pi_0(Y) \longrightarrow \pi_0(X +_A Y)$$
that is surjective. When trying to show injectivity, however, we encounter the problem of factoring paths into the pushout $X +_A Y$ over X or Y. Therefore, we will only consider the situation in which this is easily done: particularly 'nice' pushouts.

Theorem 6.2.12 (Mayer–Vietoris for Path Components)
Let X be the union of the open subsets U and V. Then the commutative diagram

$$\begin{array}{ccc} \pi_0(U \cap V) & \longrightarrow & \pi_0(V) \\ \downarrow & & \downarrow \\ \pi_0(U) & \longrightarrow & \pi_0(U \cup V) \end{array}$$

is a pushout (in the category of sets).

Proof. It suffices to show that the map
$$\pi_0(U) +_{\pi_0(U \cap V)} \pi_0(V) \longrightarrow \pi_0(U \cup V)$$
is bijective. It is certainly surjective. Let now γ be a path in X, connecting x_0 with x_1. We need to show that x_0 and x_1 also represent the same element of $\pi_0(U) +_{\pi_0(U \cap V)} \pi_0(V)$. The two open subsets $\gamma^{-1}(U)$ and $\gamma^{-1}(V)$ cover I. Because I is a compact metric space, there is, therefore, a Lebesgue number ε (see

Sect. 4.1) so that every ε-neighbourhood lies in $\gamma^{-1}(U)$ or in $\gamma^{-1}(V)$. There is then also an n, so that the intervals $[(k-1)/n, k/n]$ for $k = 1, \ldots, n$ each are mapped entirely to U or entirely to V. It is now sufficient to show that $\gamma((k-1)/n)$ represents the same element as $\gamma(k/n)$. In other words, it can be assumed that γ is entirely in U or entirely in V. In this case, the assertion is clear. □

The theorem about the sum is the special case $U \cap V = \emptyset$.

Supplement

Local Path Connectivity A topological space is called *locally path-connected*, if every neighbourhood contains a path-connected neighbourhood. For example, the space

$$X = (\{1/n \mid n \in \mathbb{N}\} \times \mathbb{R}) \cup (\mathbb{R} \times \{0\}) \cup (\{0\} \times \mathbb{R})$$

is path-connected but not locally path-connected. Every open subset of an Euclidean space is locally path-connected. Thus, every topological space for which there is a neighbourhood for each point that is homeomorphic to an open subset of $\mathbb{R}^n$ is locally path-connected.

Locally path-connected spaces that are connected are path-connected. This can be seen as follows. Let $[a]$ be the path component of a, i.e., the set of all points in X that can be connected with a. It suffices to show that $[a]$ is open because its complement, the union of the other path components, is also open. Because X is connected, the subset $[a]$ then coincides with X. For $b \in [a]$ choose a path-connected neighbourhood U. This lies in $[b]$ and thus also in $[a] = [b]$.

Exercises

Exercise 100 Path-Connected or Not?
Decide whether the following statements are true or false:

(1) Every continuous image of a path-connected space is again path-connected.
(2) Products and pullbacks of path-connected spaces are path-connected.
(3) Sums and pushouts of path-connected spaces are path-connected.
(4) The closure of a path-connected subspace is path-connected.
(5) The union of path-connected subspaces that intersect pairwise is path-connected.

Exercise 101 Faithful
Show that every functor between categories maps isomorphisms to isomorphisms.

6.3 The Concept of Homotopy

If X and Y are two topological spaces, we have endowed the set of continuous maps between them with the compact-open topology and denoted it by $\mathrm{Hom}(X, Y)$. If this space seems too complicated, we can sooner study the set $\pi_0 \mathrm{Hom}(X, Y)$ of its path components. What does this mean explicitly? A point in $\mathrm{Hom}(X, Y)$ is a continuous map from X to Y. A path in $\mathrm{Hom}(X, Y)$ thus connects one continuous map with another. If X is locally compact, then the continuity of a path $I \to \mathrm{Hom}(X, Y)$ is equivalent to the continuity of the adjoint map $I \times X \to Y$. This motivates the following definition.

Definition 6.3.1

A *homotopy* is a continuous map

$$H : I \times X \longrightarrow Y.$$

Every such homotopy provides, by adjunction, a path $H^\#$ in $\mathrm{Hom}(X, Y)$, and we then also speak of a homotopy from $H^\#(0)$ to $H^\#(1)$. Continuous maps $f, g : X \to Y$ are called *homotopic* to each other, in symbols $f \simeq g$, if there is a homotopy H with $H^\#(0) = f$ and $H^\#(1) = g$. In detail, this means

$$H(0, x) = f(x)$$
$$H(1, x) = g(x)$$

for all $x \in X$ (see Fig. 6.3).

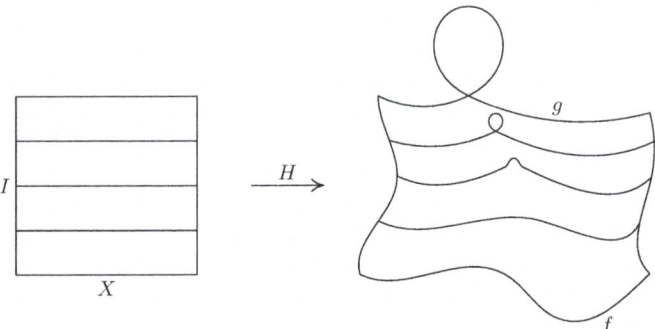

Fig. 6.3 A homotopy

Example 6.3.2
It is worth considering a seemingly tautological example first. Every element t from I defines a continuous map
$$i_t \colon X \to I \times X, \ x \longmapsto (t, x).$$
If we imagine $I \times X$ as a cylinder, then i_0 is thus the inclusion of the base and i_1 the inclusion of the lid. The identity
$$\mathrm{id} \colon I \times X \longrightarrow I \times X$$
can be understood as a homotopy from i_0 to i_1. More generally, every homotopy $H \colon I \times X \to Y$ is a homotopy from Hi_0 to Hi_1.

Example 6.3.3
Any two continuous functions f and g from X to $\mathbb{R}$ are homotopic to each other through straight lines:
$$H(t, x) = (1-t)f(x) + tg(x).$$
The same applies to continuous functions with a convex subset of $\mathbb{R}^n$ as the range (see Fig. 6.4).

Theorem 6.3.4
Homotopy defines an equivalence relation on $\mathrm{Hom}(X, Y)$.

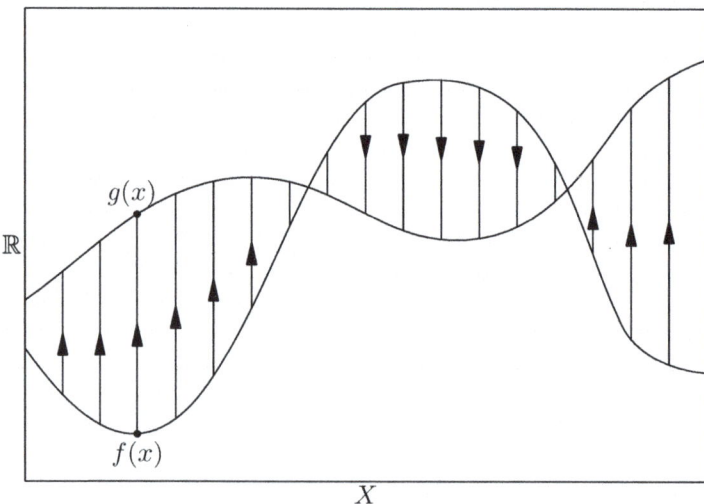

Fig. 6.4 Any two continuous functions to $\mathbb{R}$ are homotopic

6.3 The Concept of Homotopy

Proof. This follows almost immediately from the fact that connectivity defines an equivalence relation. It only needs to be verified that even the adjoints of the constructions used are continuous. For example, if f is homotopic to g by means of H and g is homotopic to h by means of H', then f is also homotopic to h by means of

$$H'(t,x) = \begin{cases} H(2t, x) & \text{for } t \leqslant 1/2 \\ H'(2t-1, x) & \text{for } t \geqslant 1/2. \end{cases}$$

The symmetry follows from the homotopy

$$(t, x) \longmapsto H(1-t, x)$$

and the reflexivity from

$$(t, x) \longmapsto f(x).$$

□

Definition 6.3.5

The set of homotopy classes of continuous maps $X \to Y$ is denoted by

$$[X, Y],$$

and $[f]$ denotes the homotopy class of a continuous map f. Two maps $f, g: X \to Y$ are thus homotopic if $[f] = [g]$ holds. A map is called *null-homotopic* if it is homotopic to a constant map.

Examples 6.3.6

If $\star$ is a singleton space, then

$$[\star, X] = \pi_0 \mathrm{Hom}(\star, X) \cong \pi_0 X.$$

This means that two points from X, considered as continuous maps $\star \to X$, are homotopic if and only if the two points are connectable. Furthermore, we have

$$[S^1, X] = \pi_0 \mathrm{Hom}(S^1, X) = \pi_0 L X,$$

and soon it will become clear why $[I, X] = \pi_0 \mathrm{Hom}(I, X) = \pi_0 P X$ is not so interesting.

Theorem 6.3.7

If $f, f': X \to Y$ are homotopic and $g, g': Y \to Z$ are homotopic, then their compositions $gf, g'f': X \to Z$ are also homotopic.

Proof. If H is a homotopy from f to f', then gH is a homotopy from gf to gf'. If K is a homotopy from g to g', then $K(\mathrm{id}_I \times f')$ is a homotopy from gf' to $g'f'$. Overall, this results in $gf \simeq gf' \simeq g'f'$. □

The homotopy relation is, therefore, compatible with the composition. It now allows us to define a category in which the continuous maps are replaced by their homotopy classes.

Definition 6.3.8

The objects of the *homotopy category* **HoTop** of topological spaces are the same as those of the usual category of topological spaces, i.e., the topological spaces themselves. But the set of morphisms from X to Y in the homotopy category is the set $[X, Y]$ of homotopy classes of continuous maps $X \to Y$. The composition is given by $[g][f] = [gf]$. The preceding theorem implies that this is well-defined. It is then also obviously associative. The identity of a space is the homotopy class of the identity, so we have $\mathrm{id} = [\mathrm{id}]$. Neutrality is then again obvious.

By construction, there is a functor

$$\mathbf{Top} \longrightarrow \mathbf{HoTop}$$

that is the identity between the objects. On morphisms it is given by the maps

$$\mathrm{Hom}(X, Y) \to [X, Y], \quad f \mapsto [f]$$

that pass to homotopy classes.

Definition 6.3.9

A continuous map $f: X \to Y$ is a *homotopy equivalence* if $[f]$ is an isomorphism in the homotopy category. This means concretely that there is a map $g: Y \to X$ such that fg and gf are homotopic to the corresponding identities. The two spaces X and Y are then called *homotopy equivalent*. A space is called *contractible* if it is homotopy equivalent to a point $\star$.

Examples 6.3.10
Homeomorphic topological spaces are automatically homotopy equivalent. The converse of this statement is wrong. The map

$$\mathbb{R}^n \longrightarrow 0$$

is a homotopy equivalence. Any map $0 \to \mathbb{R}^n$ is homotopy inverse to it, but for the inclusion, this is particularly easy to prove: The composition

$$0 \to \mathbb{R}^n \to 0$$

6.3 The Concept of Homotopy

Fig. 6.5 A star-shaped set

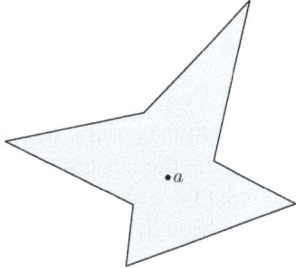

is trivially the identity and

$$I \times \mathbb{R}^n \longrightarrow \mathbb{R}^n, \ (t, x) \longmapsto tx$$

is a homotopy from the inclusion to the identity. More generally, if A is a *star-shaped* subspace of $\mathbb{R}^n$, i.e., there is a point $a \in A$ for which the straight line to every other point of A lies in A, then A is contractible (see Fig. 6.5). A null-homotopy of the identity is given by

$$H(t, x) = tx + (1-t)a$$

In particular, all convex subspaces of $\mathbb{R}^n$ are contractible.
Another example of a homotopy equivalence is the map

$$\mathbb{R}^n \setminus 0 \longrightarrow S^{n-1}, \ x \longmapsto x/\|x\|.$$

The inclusion $S^{n-1} \to \mathbb{R}^n \setminus 0$ is a homotopy inverse to it. The composition

$$S^{n-1} \to \mathbb{R}^n \setminus 0 \to S^{n-1}$$

is the identity and

$$I \times \mathbb{R}^n \setminus 0 \longrightarrow \mathbb{R}^n \setminus 0, \ (t, x) \longmapsto (1-t)x/\|x\| + tx$$

is a homotopy from the other composition to the identity.
A slightly different example is the path space PX, which is homotopy equivalent to X. For this, consider the inclusion e of X into PX through the constant paths and the map s of PX to X that assigns the starting point to each path. Then se is the identity, and

$$I \times PX \longrightarrow PX, \ (s, \gamma) \longmapsto (t \mapsto \gamma(st))$$

is a homotopy from es to the identity.

Theorem 6.3.11
If $f, g \colon X \to Y$ are homotopic, then $\pi_0(f), \pi_0(g) \colon \pi_0(X) \to \pi_0(Y)$ are equal.

Proof. This follows immediately from the fact that $\pi_0(f)$ is the map

$$[f]_* \colon [\star, X] \longrightarrow [\star, Y], \quad [x] \longmapsto [f(x)].$$

This obviously only depends on the homotopy class of f. However, we will present another argument that illustrates other important techniques. We first notice that $\pi_0(I)$ is a singleton. The map induced by the projection $\mathrm{pr}_X \colon I \times X \to X$ thus factors through bijections

$$\pi_0(I \times X) \cong \pi_0(I) \times \pi_0(X) \cong \pi_0(X),$$

and is, therefore, itself bijective. For each t from I the composition

$$X \xrightarrow{i_t} I \times X \xrightarrow{\mathrm{pr}_X} X$$

is always the identity. Applying π_0 and exploiting the functoriality implies that $\pi_0(i_t)$ does not depend on t. In particular, we have $\pi_0(i_0) = \pi_0(i_1)$. If finally H is a homotopy from f to g, then $f = Hi_0$ and $g = Hi_1$. But then $\pi_0(f) = \pi_0(g)$ follows from the functoriality of π_0. □

The previous theorem can also be interpreted as follows. The path component functor π_0 from the category of topological spaces into the category of sets factors through the homotopy category.

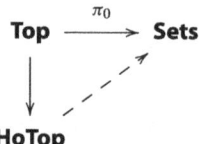

In other words, the set of path components of a topological space is a homotopy invariant.

Exercises

Exercise 102 Null-Homotopic Among Themselves
Any two null-homotopic maps $f, g \colon X \to Y$ are homotopic if and only if the space Y is path-connected.

Exercise 103 Dots
A topological space X is contractible if and only if the identity id_X on X is null-homotopic.

Exercise 104 To Be Continued?
A continuous map $f\colon S^{n-1} \to X$ is null-homotopic if and only if there is a continuous map $g\colon D^n \to X$ with $f = g|S^{n-1}$.

Exercise 105 Sphere Complements
If $m \leqslant n$, then S^m can be considered canonically as a subspace of S^n. The complement is homotopy equivalent to a sphere. (There are knotted embeddings $S^1 \to S^3$ whose complements are not homotopy equivalent to a sphere, only stably.)

Exercise 106 Linear Algebra
The inclusion $SO(2) \subseteq SL(2, \mathbb{R})$ is a homotopy equivalence.

Exercise 107 Hyperplane Complements
The j-th hyperplane in $\mathbb{R}^n$ is defined as the linear subspace $H_j = \{x \in \mathbb{R}^n \mid x_j = 0\}$ for any $j = 1, \ldots, n$. Then for any k-element subset $K \subseteq \{1, \ldots, n\}$, the complement of

$$\bigcup_{k \in K} H_k$$

in $\mathbb{R}^n$ is homotopy equivalent to a discrete set with 2^k elements. What happens when $\mathbb{R}$ is replaced by $\mathbb{C}$?

6.4 Self-maps of the Circle

After all the new terminology we have introduced in this chapter, we will now see how we can understand a first interesting example: continuous maps from the circle S^1 into itself, initially only up to homotopy, i.e., the set

$$[S^1, S^1].$$

First, we can address a more general problem, namely the sets $[X, S^1]$ for arbitrary topological spaces X. How can we even specify maps $X \to S^1$? The easiest way is perhaps, if we can find a function $X \to \mathbb{R}$ on X so that the desired map to S^1 then appears as composition with the exponential map $p\colon \mathbb{R} \to S^1$, $t \mapsto \exp(2\pi i t)$:

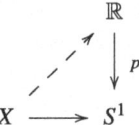

The dashed arrow is then also called a *lift* of the map along p (see Fig. 6.6).

But how can we find a lift, if at all? This is quite simple if the image does not hit a point. Because the pre-image of $S^1 \setminus z$ under p is the complement of the set $\{t + n \mid n \in \mathbb{Z}\}$, where t is a pre-image of z. This pre-image is, therefore, the disjoint union of intervals that are mapped by p homeomorphically onto $S^1 \setminus z$. The inverses of these homeomorphisms can then be used to lift the map. But does it work

Fig. 6.6 The exponential map

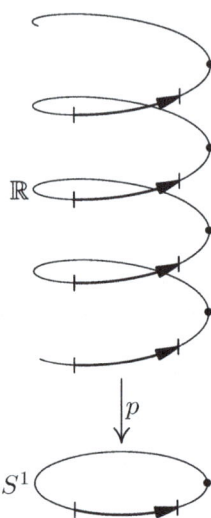

in general? This is rather unlikely, as any map that can be lifted is automatically null-homotopic because $\mathbb{R}$ is contractible. However, if it is null-homotopic, we can homotope it so that it does not hit a point because a constant map only hits one point. So, in this case, we have legitimate hopes of being able to lift the map itself. To do this, however, we must answer whether we can lift a map if we can already lift a homotopic map.

Theorem 6.4.1 (Lifting Theorem)
Let X be a connected topological space with a continuous function $f : X \to \mathbb{R}$. Let $H : I \times X \to S^1$ be a homotopy with start pf. Then there is a unique homotopy $F : I \times X \to \mathbb{R}$ with start f and $pF = H$.

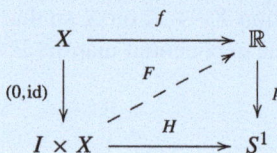

Proof. (Uniqueness.) If F and F' are two such maps, then $p(F - F')$ is equal to $H/H = 1$, in particular constant. Then the image of $F - F'$ is in $\mathbb{Z}$, so it is also constant since $I \times X$ is connected. And the constant is 0, because F and F' agree at the beginning. So $F = F'$ holds.

(Existence.) First, we consider the case where X is a point t in $\mathbb{R}$. Then H is a path γ in S^1 with $\gamma(0) = p(t)$. According to the Lebesgue lemma, there is an n such that γ maps each interval of $[(k-1)/n, k/n]$ into a set of the form $S^1 \setminus z$. From

6.4 Self-maps of the Circle

the discussion preceding this theorem, it follows that the restriction of γ to $[0, 1/n]$ has (exactly) one lift with start t. The endpoint of this lifting is then used to continue the lifting over $[1/n, 2/n]$ and so on. In this way, we obtain a lifting of γ over the whole interval $[0, 1]$. In the general case, it is now clear how F should look as a map. If x is a point from X, then the restriction of F on $I \times \{x\}$ must be the lifting of the restriction of H on $I \times \{x\}$ to the beginning $f(x)$. Only the continuity of this map is left to be shown. The argument will be postponed to later when this theorem is proven in greater generality (see Theorem 8.2.2). □

After these preparations, we can start to classify continuous self-maps of the circle—for now up to homotopy. If $f : S^1 \to S^1$ is a continuous self-map, it may not be null-homotopic. The strategy considered above, therefore, does not always work without further ado. But the composition of f with the identification $q : I \to S^1$, $t \mapsto \exp(2\pi i t)$ yields a path fq in S^1, which due to the contractibility of I is automatically null-homotopic. This composition can then be lifted along p to a path $\tilde{f}$ in $\mathbb{R}$.

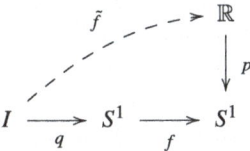

More precisely, for each pre-image of $f(1)$ under p, there is a unique lift $\tilde{f}$ with this initial value. From the lifting theorem, it also follows that any two such lifts with different initial values differ only by a constant value. If we add integer constants to $\tilde{f}$, we get all lifts. In particular, the difference $\tilde{f}(1) - \tilde{f}(0)$ does not depend on the choice of the initial value. This difference is an integer because $\tilde{f}(1)$ and $\tilde{f}(0)$ have the same image. Thus, each continuous map $f : S^1 \to S^1$ has been assigned an integer.

Definition 6.4.2

The integer $\tilde{f}(1) - \tilde{f}(0)$ is called the *mapping degree* $\deg(f)$ of f.

Example 6.4.3
Consider the continuous maps
$$e_n : S^1 \longrightarrow S^1, \; z \mapsto z^n$$
for n from $\mathbb{Z}$. A lift of this is $\tilde{e}_n(t) = nt$ because
$$p\tilde{e}_n = \exp(2\pi i n t) = \exp(2\pi i t)^n = e_n q(t).$$
The beginning of the lift is 0, the end is n. Therefore, this map has the mapping degree n.

Theorem 6.4.4
The mapping degree is a homotopy invariant. If f and g are homotopic maps, then $\deg(f) = \deg(g)$ holds.

Proof. Let $\tilde{f}$ be a lift of fq and $H \colon I \times S^1 \to S^1$ a homotopy from f to g. Then there is (exactly) one lift F of $H(\mathrm{id}_I \times q)$ with the beginning $\tilde{f}$.

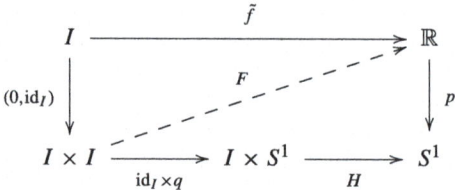

The end of F is a lift of gq and can thus be used to calculate the degree of g:

$$\deg(g) = F(1,1) - F(1,0).$$

The paths $s \mapsto H(\mathrm{id}_I \times q)(s, 0)$ and $s \mapsto H(\mathrm{id}_I \times q)(s, 1)$ from $f(1)$ to $g(1)$ in S^1 agree because $q(0) = q(1)$. Then $s \mapsto F(s, 0)$ and $s \mapsto F(s, 1)$ are two lifts of the same path. Their difference is therefore constant. This means that

$$F(s, 1) - F(s, 0)$$

does not depend on s. Hence,

$$\deg(g) = F(1,1) - F(1,0) = F(0,1) - F(0,0) = \deg(f),$$

as desired. □

The mapping degree thus induces a map

$$\deg \colon [S^1, S^1] \longrightarrow \mathbb{Z}, \; [f] \longmapsto \deg(f).$$

The above examples show that this map is surjective.

Corollary 6.4.5
The space S^1 is not contractible. In particular, the circle S^1 is not a retract of the disk D^2.

6.4 Self-maps of the Circle

Fig. 6.7 A retraction of the disk to its edge?

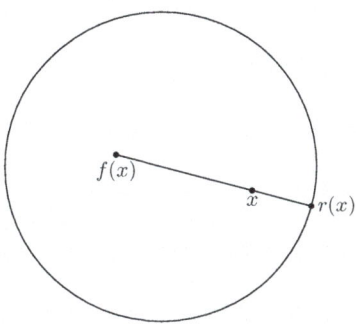

Proof. Otherwise, the set $[S^1, S^1]$ would have a unique element, so it could not map surjectively onto $\mathbb{Z}$. For the second assertion, assume that $r: D^2 \to S^1$ is a retraction map, so it coincides with the identity map on S^1. Then choose a null-homotopy of the identity of D^2, combine it with r and restrict it to $I \times S^1$. This would result in a contraction of the circle, which is impossible. □

> **Corollary 6.4.6** (Brouwer's Fixed Point Theorem for the Disk)
> If $f: D^2 \to D^2$ is continuous, then there exists a point $x \in D^2$ with $f(x) = x$.

Proof. Suppose f is a continuous self-map of the disk without a fixed point. Then the connecting line from $f(x)$ to x intersects the circle in a point $r(x) \in S^1$. In this way, a continuous retraction r from the disk to the circle line has been created (see Fig. 6.7). □

▶ **Remark 6.4.7** We say that a topological space X has the *fixed point property* if every continuous self-map $f: X \to X$ has at least one fixed point. Brouwer's theorem thus states that the closed disk D^2 has the fixed point property. The same applies to any closed interval (according to the intermediate value theorem from the introductory course in analysis). But what happens if the self-map is continuously varied? Can we then always select a fixed point continuously? These questions are answered in [Szy14].

Is the above assignment of the mapping degree also injective? In other words, are two continuous maps with the same degree also homotopic? To answer this, it is worth enriching the situation with some structure. The space $LS^1 = \mathrm{Hom}(S^1, S^1)$ is a topological group under pointwise multiplication. (That was an exercise.) The set $[S^1, S^1]$ of path components then receives an induced group structure. For example, we have $e_m \cdot e_n = e_{m+n}$, and this raises the question of whether in general the mapping degree is compatible with the group structures.

Theorem 6.4.8
The map

$$\deg\colon [S^1, S^1] \longrightarrow \mathbb{Z}$$

is a group homomorphism. We have

$$\deg(f \cdot g) = \deg(f) + \deg(g)$$

for any two continuous maps $f, g\colon S^1 \to S^1$.

Proof. If $\tilde{f}$ is a lift of fq and $\tilde{g}$ is a lift of gq, then $\tilde{f} + \tilde{g}$ is a lift of $(f \cdot g)q$. From the calculation

$$\begin{aligned}\deg(f \cdot g) &= (\tilde{f} + \tilde{g})(1) - (\tilde{f} + \tilde{g})(0) \\ &= (\tilde{f}(1) - \tilde{f}(0)) + (\tilde{g}(1) - \tilde{g}(0)) \\ &= \deg(f) + \deg(g)\end{aligned}$$

the assertion follows. □

Theorem 6.4.9
The map induced by the degree

$$\deg\colon [S^1, S^1] \xrightarrow{\cong} \mathbb{Z}$$

is bijective, thus an isomorphism of groups.

Proof. Since it is a group homomorphism, it is sufficient to show that the kernel is trivial. So let $f\colon S^1 \to S^1$ be a map of degree 0. For each lift $\tilde{f}$, the starting point and endpoint coincide. Thus, the map $\tilde{f}$ defines a loop in $\mathbb{R}$ that factorises f.

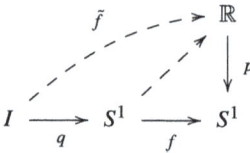

Thus, the map f is null-homotopic. □

6.4 Self-maps of the Circle

Now that the path component set of $LS^1 = \text{Hom}(S^1, S^1)$ is understood, we can try to understand the individual components. Because the space admits a compatible group structure, the components are all homeomorphic. It is, therefore, sufficient to look at the component of the neutral element. The neutral element is the constant function with value 1. This component consists of all continuous maps of degree 0.

Theorem 6.4.10
The map $\deg \colon \text{Hom}(S^1, S^1) \to \mathbb{Z}$ *is continuous.*

Proof. It suffices to show that the path components, i.e., the subsets of self-maps of a fixed degree, are open. This only needs to be verified for the component of the neutral element. Because this component is a subgroup, it is enough to find an open neighbourhood of the neutral element that is entirely within this component. Such a neighbourhood is $M(S^1, S^1 \setminus -1)$ because all maps therein are null-homotopic. □

The map
$$\text{Hom}(S^1, S^1) \longrightarrow \mathbb{Z} \times S^1, \ f \longmapsto (\deg(f), f(1))$$
is a continuous surjective group homomorphism. Let H be its kernel. It consists of all continuous self-maps of the circle that are null-homotopic and fix 1. Then
$$\text{Hom}(S^1, S^1) \to \mathbb{Z} \times S^1 \times H, \ f \mapsto \left(\deg(f), f(1), f \cdot f(1)^{-1} \cdot e_{-\deg(f)}\right)$$
is an isomorphism of topological groups with inverse
$$\mathbb{Z} \times S^1 \times H \longrightarrow \text{Hom}(S^1, S^1), \ (n, z, f) \longmapsto e_n \cdot z \cdot f.$$

Theorem 6.4.11
The space H is contractible.

Proof. The map p_* induces a continuous bijection from the (contractible) vector space V of functions $S^1 \to \mathbb{R}$ with $1 \mapsto 0$ to H. It suffices to show that this map is open and, therefore, a homeomorphism. Since it is a group homomorphism, it is only necessary to show that open neighbourhoods of the neutral element are mapped to open neighbourhoods of the neutral element. For this, it is sufficient to consider sets of the form $M(K, U)$, where $K \subseteq S^1$ is compact and $U \subseteq {]-1/2, +1/2[}$ is an open neighbourhood of 0. Since p is open, then $p(U) \subseteq S^1 \setminus -1$ is open in S^1. We have $p_* M(K, U) = M(K, p(U))$ due to the initial conditions, and that is open. □

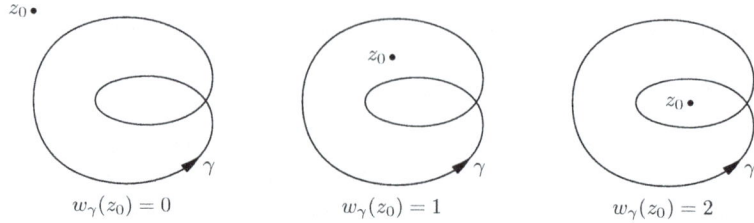

Fig. 6.8 The winding numbers of some planar curves

Corollary 6.4.12
The space $\mathrm{Hom}(S^1, S^1)$ is homotopy equivalent to $\mathbb{Z} \times S^1$.

Supplement

Winding and Rotation Numbers In complex analysis, the *winding number* of a smooth loop in $\mathbb{C}$ with respect to a point $z_0 \in \mathbb{C}$ that is not in the image of γ, is defined as

$$w_\gamma(z_0) = \frac{1}{2\pi i} \oint_\gamma \frac{dz}{z - z_0}.$$

For example, the paths e_n have the winding number n around the origin (see Fig. 6.8). It can be shown that the winding number only depends on the homotopy class

$$[\gamma] \in [S^1, \mathbb{C} \setminus z_0].$$

Thus, the results of this section imply the equality

$$w_\gamma(z_0) = \deg\left(\frac{\gamma - z_0}{\|\gamma - z_0\|} : S^1 \longrightarrow S^1 \right).$$

The winding number is to be distinguished from the rotation number. A continuously differentiable planar curve $\gamma \colon S^1 \to \mathbb{R}^2$ whose derivative $\gamma' \colon S^1 \to \mathbb{R}^2$ never vanishes is called an *immersion*. The *rotation number* of an immersion is the winding number of its derivative around zero [Whi37].

Exercises

Exercise 108 The Degree of Compositions
If f and g are two continuous self-maps of the circle, then their degrees satisfy $\deg(g \circ f) = \deg(g) \cdot \deg(f)$.

Exercise 109 A Small Fixed Point Theorem
Every continuous self-map f of the circle with $\deg(f) \neq 1$ has a fixed point.

Exercise 110 Another Pullback
The commutative diagram

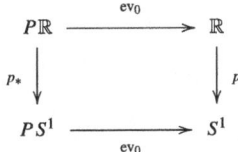

is a pullback. (Here is a hint. To find a map $PS^1 \times_{S^1} \mathbb{R} \to P\mathbb{R}$, the homotopy lifting theorem can be applied to $X = PS^1 \times_{S^1} \mathbb{R}$ and $f = \mathrm{pr}_2$ and $H(s, \gamma, t) = \gamma(s)$.)

Fundamental Groups 7

In this chapter, we define the fundamental group of a (pointed) space as the set of path components of the corresponding loop space and study its properties. Initially, we can imagine that maps from the circle S^1 into a topological space X (that map 1 to a fixed point x) always have a 'generalised mapping degree'. However, it does not take values in $\mathbb{Z}$ but in the fundamental group $\pi_1(X, x)$. We will develop techniques for calculating these fundamental groups by characterising their behaviour with respect to coverings. Subsequently, we illustrate the results in the case of surfaces.

7.1 Fundamental Groupoids

On the way from a space X to the set $\pi_0(X)$ of its path components, we identify points that can be connected by a path. It is often helpful with such identifications not only to know whether two points are identified but in how many ways we can identify them.

Definition 7.1.1

For a topological space X, the category $\Pi(X)$ is defined as follows: the object set of $\Pi(X)$ is the underlying set of X. An object of $\Pi(X)$ is thus the same as a point of X. The set of morphisms from an object x_0 to an object x_1 is

$$\mathrm{Mor}_{\Pi(X)}(x_0, x_1) = \pi_0 PX(x_0, x_1).$$

Each morphism $x_0 \to x_1$ is thus represented by a path γ from x_0 to x_1. Another path γ' from x_0 to x_1 represents the same morphism if there is a homotopy

$$H: I \times I \longrightarrow X$$

Fig. 7.1 A bounded homotopy

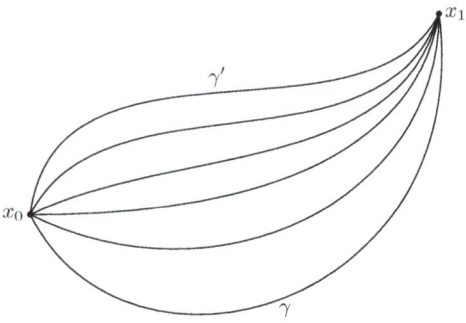

with $H(0, t) = \gamma(t)$ and $H(1, t) = \gamma'(t)$ and $H(s, 0) = x_0$ and $H(s, 1) = x_1$ for all $s, t \in I$. Thus, the starting point and endpoint remain fixed during homotopy. This is also referred to as a *bounded homotopy* (Fig. 7.1).

The composition in $\Pi(X)$ is given by the concatenation of paths. This is not associative because the speed changes when concatenating. We may wonder, however, why the homotopy class of a path (with fixed start and end points) does not depend on the parameterisation. We will prove this for the universal example first.

Theorem 7.1.2
The space $PI(0, 1)$ is contractible.

Proof. Consider the map $\star \to PI(0, 1)$ that maps the point to id_I. It is a homotopy equivalence because

$$I \times PI(0, 1) \longrightarrow PI(0, 1), \quad (s, \gamma) \longmapsto (t \mapsto st + (1-s)\gamma(t))$$

is a homotopy from the identity of $PI(0, 1)$ to the constant map with value id_I. □

Given a path γ in X and a reparameterisation φ of the interval, it follows that $\gamma = \gamma \mathrm{id}$ and $\gamma\varphi$ are homotopic in X and hence provide the same morphism in $\Pi(X)$. Thus, the associativity axiom for the composition is fulfilled. The identity of an object x is the constant path $e(x)$ at x. It does not behave neutrally with respect to the composition of paths, but it does so after taking homotopy classes. Again, this follows from the reparameterisation argument (Fig. 7.2).

Thus, we have explained all data of the category $\Pi(X)$ and verified all axioms. The category $\Pi(X)$ is called the *fundamental groupoid* of X.

The terminology can be justified as follows.

7.1 Fundamental Groupoids

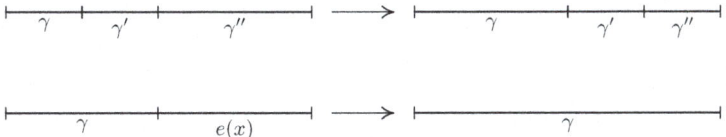

Fig. 7.2 Reparameterisations of paths

Definition 7.1.3

A *groupoid* is a category in which every morphism is an isomorphism. The category of small groupoids and functors between them is denoted by **Grpd**.

Examples 7.1.4

Examples of groupoids abound in mathematics. Every category provides a groupoid if all morphisms that are not isomorphisms are removed. Every group, understood as a category with one object and the group as endomorphisms, is a groupoid. In this sense, groupoids are 'groups with multiple objects'. An equivalence relation on a set yields a groupoid with this set of objects. The set of morphisms between two elements is single-element if the elements are equivalent and empty otherwise. In this sense, groupoids are equivalence relations where we not only remember *if* but also possibly *how* two objects are equivalent, i.e., by which isomorphisms. Suppose G is a group that acts on a set X. Then we may form the *transport groupoid*. Its set of objects is X, and the set of morphisms is $G \times X$. More precisely, the pair (g, x) is a morphism from x to gx. The concatenation, the identity and the inverse are given by

$$(h, y) \circ (g, x) = (hg, x)$$
$$\mathrm{id}_x = (1, x)$$
$$(g, x)^{-1} = (g^{-1}, gx).$$

In topology, the following theorem gives the most important examples of groupoids.

Theorem 7.1.5

The fundamental groupoid is a groupoid.

Proof. Paths can be inverted up to bounded homotopy: if γ is a path from x_0 to x_1 then the path γ^- runs in the opposite direction from x_1 to x_0. The composites $\gamma^-\gamma$ and $\gamma\gamma^-$ are not constant in general (at x_0 or x_1) but they are bounded homotopic to the constant ones. To see this, we do not go to the end of the path but wait in between at a point. During the homotopy, we always wait longer at a point when getting closer to the starting point until the path is finally constant (Fig. 7.3). □

Now that a category $\Pi(X)$ is defined, the question arises immediately: what isomorphism classes exist, and what do their automorphism groups look like? The former is simple.

Fig. 7.3 A homotopy from $\gamma\gamma^-$ to the constant path

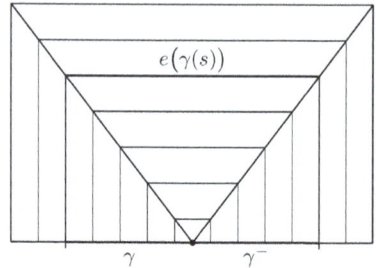

▶ **Remark 7.1.6** The set of isomorphism classes of $\Pi(X)$ is equal to the set $\pi_0(X)$ of path components of X.

The set of isomorphism classes of objects, therefore, provides nothing new. The automorphism groups do:

Definition 7.1.7

If x is a point from X, then $\mathrm{Aut}_{\Pi(X)}(x)$ is a group, the *fundamental group* of X at x. It is denoted by

$$\pi_1(X, x).$$

Of course, this group depends on the choice of the point x in X.

By definition,

$$\pi_1(X, x) = \mathrm{Aut}_{\Pi(X)}(x) = \mathrm{Mor}_{\Pi(X)}(x, x) = \pi_0 PX(x, x) = \pi_0 \Omega(X, x).$$

Hence, we have shown that $\pi_0\Omega(X, x)$ always is a group. There is a distinguished point in the space $\Omega(X, x)$, namely the constant path at x. We can iterate the construction:

$$\Omega^2(X, x) = \Omega(\Omega(X, x), e(x))$$

According to the exponential law, an element in $\Omega^2(X, x)$ corresponds to a map from the square $I^2 = I \times I$ to X that maps the edges ∂I^2 to x. We may continue and consider the space $\Omega^n(X, x)$ for larger n. The sets of path components are groups for the same reason as for $n = 1$.

Definition 7.1.8

The *higher homotopy groups* of X at x are defined by

$$\pi_n(X, x) = \pi_0 \Omega^n(X, x).$$

7.1 Fundamental Groupoids

The higher homotopy groups also depend on the choice of x. To round off the notation, we sometimes also define $\pi_0(X, x)$, namely as a set $\pi_0(X)$ with a distinguished element, namely the component of x; this is not a group, in general.

The dependence of the fundamental group on the base point is well under control. If γ is a path from x to x' in X, then

$$\pi_1(X, x) \longrightarrow \pi_1(X, x'), \quad \omega \longmapsto \gamma\omega\gamma^-$$

is an isomorphism of groups. In this way, the assignment

$$x \mapsto \pi_1(X, x)$$

becomes a functor

$$\Pi(X) \longrightarrow \mathbf{Grp}.$$

The assignment

$$\Pi \colon \mathbf{Top} \longrightarrow \mathbf{Grpd},$$

that assigns to each topological space X its fundamental groupoid is also functorial. This means that a continuous map $f \colon X \to Y$ induces the functor $\Pi(f) \colon \Pi(X) \to \Pi(Y)$ by

$$\Pi(f)[\gamma] = [f\gamma].$$

In Theorem 6.3.11, it was already proven that the functor π_0 is homotopy invariant. What about Π (and π_1)? For trivial reasons, if $f, g \colon X \to Y$ are continuous maps with the property that $\Pi(f)$ and $\Pi(g)$ coincide then f and g are equal. This is because Π keeps information about the objects, which are the points. Similarly, a continuous map is a bijection if it induces an isomorphism between the fundamental groupoids. Hence, we need to look for a different relationship between the induced functors of homotopic maps. The following terms in the language of categories will prove useful.

Definition 7.1.9

Let F and G be functors from $\mathcal{C}$ to $\mathcal{D}$. A *natural transformation* Φ from F to G is an assignment that gives a morphism in $\mathcal{D}$,

$$\Phi_X \colon F(X) \longrightarrow G(X)$$

for each object X of $\mathcal{C}$ such that for all morphisms $f\colon X \to Y$ in $\mathcal{C}$ the diagram

$$\begin{array}{ccc} F(X) & \xrightarrow{\Phi_X} & G(X) \\ {\scriptstyle Ff}\downarrow & & \downarrow{\scriptstyle Gf} \\ F(Y) & \xrightarrow{\Phi_Y} & G(Y) \end{array}$$

commutes. A natural transformation is called *natural equivalence* or *natural isomorphism* if Φ_X is an isomorphism for all X.

Theorem 7.1.10
Every homotopy $f \simeq g\colon X \to Y$ induces a natural transformation

$$\Pi(f) \cong \Pi(g)\colon \Pi(X) \longrightarrow \Pi(Y)$$

of functors. Since $\Pi(Y)$ is a groupoid, it is automatically a natural equivalence.

Proof. Let $H\colon I \times X \to Y$ be the homotopy. We have to give a morphism from $\Pi(f)(x)$ to $\Pi(g)(x)$ for each object x from $\Pi(X)$. We choose for this the class of the path $t \mapsto H(t,x)$. To get a natural transformation, we have to check that the diagram

$$\begin{array}{ccc} f(x) & \xrightarrow{H(?,x)} & g(x) \\ {\scriptstyle f(\gamma)}\downarrow & & \downarrow{\scriptstyle g(\gamma)} \\ f(x') & \xrightarrow[H(?,x')]{} & g(x') \end{array}$$

in $\Pi(Y)$ commutes for each morphism $[\gamma]\colon x \to x'$ in $\Pi(X)$. This means that $g(\gamma)H(?,x)$ and $H(?,x')f(\gamma)$ are homotopic relative to the start and end point, and it follows from the fact that $H(\mathrm{id}_I \times \gamma)$ extends the map given by the above diagram from the edge of the square to the whole square. In formulas, a homotopy looks as follows:

$$(s,t) \mapsto \begin{cases} f(\gamma(2t)) & t \leqslant s/2 \\ H(2t-s, \gamma(s)) & s/2 \leqslant t \leqslant (s+1)/2 \\ g(\gamma(2t-1)) & (s+1)/2 \leqslant t. \end{cases}$$

$\square$

7.1 Fundamental Groupoids

Definition 7.1.11

A functor $F: \mathcal{C} \to \mathcal{D}$ is called an *equivalence of categories* if there is a $G: \mathcal{D} \to \mathcal{C}$ and natural equivalences from GF to the identity functor of $\mathcal{C}$ as well as from FG to the identity of $\mathcal{D}$.

Corollary 7.1.12
If f is a homotopy equivalence, then $\Pi(f)$ is an equivalence of categories.

So, the functor Π does not factorise through the homotopy category. However, it does induce a functor from the homotopy category to the category of groupoids and natural isomorphism classes of functors. Here again, the transition to isomorphism classes of objects is well-defined, and we obtain a factorisation (read: improvement) of π_0 through Π, but we will not go into more detail here. As with the path component functor, it is easy to see the (natural!) isomorphisms

$$\Pi(X \times Y) \cong \Pi(X) \times \Pi(Y)$$

and

$$\Pi(X) + \Pi(Y) \cong \Pi(X + Y).$$

The sum and product of groupoids are defined in an obvious way, and the claim above is not to be understood in such a way that we must know their constructions. Instead, it asserts that there are groupoids that fulfil the universal property. At the end of this section, we want to investigate to what extent the fundamental group determines the fundamental groupoid. We first introduce some more vocabulary.

Definition 7.1.13

A functor $F: \mathcal{C} \to \mathcal{D}$ is called *fully faithful* if the map given by F

$$\text{Mor}_{\mathcal{C}}(X, X') \longrightarrow \text{Mor}_{\mathcal{D}}(F(X), F(X'))$$

is bijective for all objects X and X' of $\mathcal{C}$. It is called *essentially surjective* if, for each object Y of $\mathcal{D}$, there is an object X of $\mathcal{C}$ such that Y is isomorphic to $F(X)$.

Theorem 7.1.14
A functor is an equivalence of categories if and only if it is fully faithful and essentially surjective.

Fig. 7.4 Equivalent groupoids

Proof. Let $F: \mathcal{C} \to \mathcal{D}$ be fully faithful and essentially surjective. Then we need to construct an inverse equivalence $G: \mathcal{D} \to \mathcal{C}$. For each object Y of $\mathcal{D}$, choose an object $G(Y)$ of $\mathcal{C}$ with $Y \cong FG(Y)$. Also, choose an isomorphism $\Phi_Y: Y \to FG(Y)$. Then G is defined on the objects of $\mathcal{D}$. Let $g: Y \to Y'$ be a morphism in $\mathcal{D}$. Then there is a unique morphism $G(g): G(Y) \to G(Y')$ that, under F, maps to the composite

$$FG(Y) \xrightarrow{\Phi_Y^{-1}} Y \xrightarrow{g} Y' \xrightarrow{\Phi_{Y'}} FG(Y').$$

This defines the functor G on morphisms. We can check that the so-defined G is functorial. Furthermore, the functor G was just defined in such a way that the isomorphisms Φ_Y form a natural transformation $\mathrm{id}_\mathcal{D} \to FG$. To construct a natural transformation $\Psi: \mathrm{id}_\mathcal{C} \to GF$, we use the morphism $\Psi_X: X \to GF(X)$ for each object X of $\mathcal{C}$. It is mapped by F to $\Phi_{F(X)}: F(X) \to FG(F(X))$. This is (!) then an isomorphism. To check that Ψ is natural, it suffices to apply F to the relevant diagram. The resulting diagram commutes due to the naturality of Φ. The other direction directly follows from the definition. □

Example 7.1.15
Figure 7.4 shows two groupoids. The left one has one object, and the right one has three objects, but all are isomorphic to each other. The automorphism groups of all four objects are trivial.
The functor from left to right, which maps the object to one of the three objects, is essentially surjective and fully faithful. According to the theorem, it is an equivalence of categories (even of groupoids). Of course, an inverse can also be explicitly given (exercise!).

By the way, according to the theorem, it is always the case that a groupoid is equivalent to any subgroupoid that arises from it by selecting a representative from each isomorphism class and removing all other objects so that every remaining morphism is an automorphism. An instance of this observation is also the following result.

Corollary 7.1.16
If X is path-connected, then for every point x from X, the inclusion

$$\pi_1(X, x) \longrightarrow \Pi(X)$$

is an equivalence of categories.

7.1 Fundamental Groupoids

Proof. We can consider the fundamental group as a category with one object. By definition of the fundamental group, the inclusion functor is fully faithful. It is essentially surjective if and only if every point of X can be connected with x, that is, if and only if X is path-connected. Thus, it is an equivalence of categories. □

Corollary 7.1.17
If $f: X \to Y$ is a homotopy equivalence then
$$\pi_1(f): \pi_1(X, x) \longrightarrow \pi_1(Y, f(x))$$
is an isomorphism of groups for all $x \in X$.

Proof. It has already been shown that $\Pi(f)$ is an equivalence of categories. Therefore, the map induced on the object x
$$\Pi(f): \mathrm{Mor}_{\Pi(X)}(x, x) \longrightarrow \mathrm{Mor}_{\Pi(Y)}(f(x), f(x))$$
is an isomorphism. □

Definition 7.1.18

A topological space X is called *simply-connected* if its fundamental groupoid is equivalent to the groupoid with a unique morphism. This holds if and only if the space is path-connected and one (and hence all) of its fundamental groups is trivial.

Examples 7.1.19
Every contractible space is simply-connected. The circle S^1 is path-connected but not simply-connected. This follows from the results of Sect. 6.4, which imply that the fundamental groups of the circle are isomorphic to $\mathbb{Z}$; see also the subsequent exercises. Further examples will follow after the next section once we have established another tool for calculating fundamental groups.

Supplement

Commutativity of Higher Homotopy Groups Unlike the fundamental group, the higher homotopy groups are always abelian groups. To see this, consider an element from $\Omega^n(X, x)$ as a continuous map

$$f: I^n/\partial I^n \cong S^n \longrightarrow X.$$

The addition of two such maps f, g is achieved by dividing the first coordinate into two halves. A homotopy between $f + g$ and $g + f$ is suggested for $n = 2$ in Fig. 7.5.

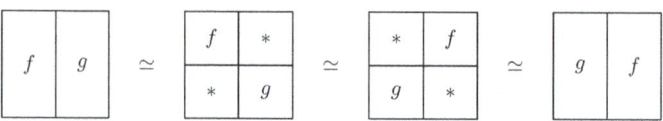

Fig. 7.5 On the commutativity of higher homotopy groups

Exercises

Exercise 111 The Fundamental Groups of the Circle
For every point z of S^1, there is an isomorphism $\varphi_z \colon \mathbb{Z} \to \pi_1(S^1, z)$ that maps the generator 1 of $\mathbb{Z}$ to the loop defined by $p_z(t) = \exp(2\pi i t)z$. For every continuous map $f \colon S^1 \to S^1$, the diagram

$$\begin{array}{ccc} \pi_1(S^1, z) & \xrightarrow{f_*} & \pi_1(S^1, f(z)) \\ \varphi_z \downarrow & & \downarrow \varphi_{f(z)} \\ \mathbb{Z} & \xrightarrow{\deg(f)} & \mathbb{Z} \end{array}$$

is commutative.

Exercise 112 A Fibre Transport Functor
The assignment that assigns to each point z from S^1 its fibre $p^{-1}(z)$ in $\mathbb{R}$ can be extended to a functor $\Pi(S^1) \to \mathbf{Sets}$. For every path $\gamma \colon z \to z'$ the induced map $p^{-1}(z) \to p^{-1}(z')$ maps a point to the endpoint of the lifting of γ at this point.

Exercise 113 Two Kinds of Multiplication
Let G be a topological group with neutral element 1. Then $\pi_1(G, 1)$ is abelian.

Exercise 114 Counterexamples
There is an embedding $f \colon X \to Y$ such that an induced homomorphism $f_* \colon \pi_1(X, x) \to \pi_1(Y, f(x))$ is not injective. There is an identification such that an induced homomorphism is not surjective.

Exercise 115 Freedom to the Loops!
Let X be a path-connected space and x be a point in it. Then the forgetful map

$$v \colon \pi_1(X, x) \longrightarrow [\, S^1, X\,]$$

is surjective. Two elements of the fundamental group have the same image if and only if they are conjugate.

Exercise 116 Loops with Ambience?
Let X be locally compact and x be a point in it. Then the evaluation at x induces a group homomorphism

$$\pi_1(\mathrm{Hom}(X, X), \mathrm{id}_X) \longrightarrow \pi_1(X, x).$$

The image lies in the centre of $\pi_1(X, x)$.

Exercise 117 Retracts Again
Let $s\colon X \to Y$ and $r\colon Y \to X$ be continuous maps with $rs = \mathrm{id}_X$, so r is a retraction with section s. For every point x of X, the group $\pi_1(Y, s(x))$ is a semi-direct product $G \rtimes \pi_1(X, x)$ for a group G.

7.2 The Seifert–van Kampen Theorem

In this section, we address the Mayer–Vietoris problem for the fundamental groupoid functor Π (and also the fundamental group functor π_1). Theorem 6.2.12 has already clarified the situation for π_0. So let X be a topological space that is covered by open sets U and V. Then the inclusions provide a pushout

$$\begin{array}{ccc} U \cap V & \xrightarrow{i_V} & V \\ {\scriptstyle i_U}\downarrow & & \downarrow{\scriptstyle j_V} \\ U & \xrightarrow{j_U} & U \cup V. \end{array}$$

Applying Π provides a commutative square, and we should ask whether this is a pushout. This term has only been introduced for sets and topological spaces. Let us generalise it as quickly as possible.

Definition 7.2.1

Let $\mathcal{C}$ be a category. A *pushout* or *co-cartesian square* in $\mathcal{C}$ is a diagram of the form

$$\begin{array}{ccc} X_0 & \xrightarrow{j_1} & X_1 \\ {\scriptstyle j_2}\downarrow & & \downarrow{\scriptstyle i_1} \\ X_2 & \xrightarrow{i_2} & X \end{array}$$

with the following universal property: for each object T of $\mathcal{C}$ and for two morphisms $f_i\colon X_i \to T$, with $i = 1, 2$, with the same target T and

$$f_1 j_1 = f_2 j_2$$

there is a unique $f\colon X \to T$ with

$$f i_k = f_k,$$

for $k = 1, 2$.

With the help of the universal property, we consider:

▶ **Remark 7.2.2** In a pushout, the object X is uniquely determined up to canonical isomorphism by

$$X_1 \xleftarrow{j_1} X_0 \xrightarrow{j_2} X_2$$

Example 7.2.3
For the category of groups and homomorphisms, we first consider the case where the top left corner G_0 is the trivial group, i.e., it consists only of the neutral element. Then the maps j_k for $k = 1, 2$ are already uniquely determined, and the universal property reduces to the *sum diagram*

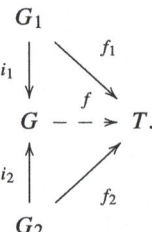

In the category of groups, a sum is also called a *free product*, even though it is not a product in the categorical sense of the word. Free products are determined up to isomorphism by their universal property, and in all cases of interest to us, they will be given to us; we do not need to know how to construct them! That said, it might instil some confidence that it is possible to construct them algebraically.

Elements in the free product are equivalence classes of finite sequences of elements in the disjoint union $G_1 + G_2$, the so-called 'words'.

$$G = G_1 * G_2 = \{(g_1, g_2, \ldots, g_n) \mid n \geq 0, g_k \in G_1 + G_2\}/\sim$$

Each word $(g_1, \ldots, g_n)$ is identified with the shortened word

$$(g_1, \ldots, g_{i-1}, g_i g_{i+1}, g_{i+2}, \ldots, g_n)$$

if the two elements g_i and g_{i+1} are in the same summand and thus a multiplication of the elements is possible. In addition, the neutral elements from G_1 or G_2 may always be shortened. The multiplication in the free product is given by the concatenation of words:

$$[g_1, \ldots, g_n][g'_1, \ldots, g'_m] = [g_1, \ldots g_n, g'_1, \ldots, g'_m].$$

The class $[\ ]$ of the empty word thus represents the neutral element. The universal property of the free product is quickly proven: the homomorphism $f: G_1 * G_2 \to T$ necessarily takes the form

$$[g_1, g_2, \ldots, g_n] \mapsto f_{v_1}(g_1) f_{v_2}(g_2) \cdots f_{v_n}(g_n),$$

where $v_k \in \{1, 2\}$ is chosen in a way that g_k is in the summand G_{v_k}. This map is well-defined because f_1 and f_2 are homomorphisms.

For arbitrary groups G_0, we proceed similarly and we define

$$U = \{[j_1(g)][j_2(g)^{-1}] \mid g \in G_0\} \subseteq G_1 * G_2.$$

7.2 The Seifert–van Kampen Theorem

If then $f_i : G_i \to Y$, for $i = 1, 2$, are two maps that satisfy $f_1 j_1 = f_2 j_2$, then U is in the kernel of the map

$$(f_1, f_2) : G_1 * G_2 \longrightarrow T.$$

The kernel of a homomorphism is always a normal subgroup. Hence, the normal subgroup $N(U)$ generated by U is in the kernel. (The generated normal subgroup $N(U)$ is U itself if U is a normal subgroup. In general, it is the intersection of all normal subgroups that contain U.) This results in a well-defined map

$$f : G = (G_1 * G_2)/N(U) \longrightarrow T.$$

The quotient group G, along with the obvious homomorphisms from G_1 and G_2, has the universal property of pushouts. The homomorphism f was constructed in such a way that it makes the two triangles commute. It is also unique because the map to the representative elements of the free product is unique.

Now, we will return to the original question. Again, the following theorem does not mean that we need to know how to construct pushouts in the category of groupoids. Rather, it ensures that there is a pushout in the assumed situation, and this is concretely realised by the fundamental groupoid of the space.

> **Theorem 7.2.4 (Seifert–van Kampen for Fundamental Groupoids)**
> Let X be a topological space that is covered by open sets U and V. Then the diagram
>
> $$\begin{array}{ccc} \Pi(U \cap V) & \xrightarrow{\Pi(i_V)} & \Pi(V) \\ {\scriptstyle \Pi(i_U)} \downarrow & & \downarrow {\scriptstyle \Pi(j_V)} \\ \Pi(U) & \xrightarrow{\Pi(j_U)} & \Pi(U \cup V) \end{array}$$
>
> is a pushout of groupoids.

Proof. It is sufficient to prove the universal property. Let Γ be any groupoid and

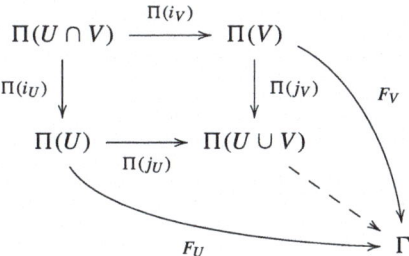

Fig. 7.6 A decomposition of a path

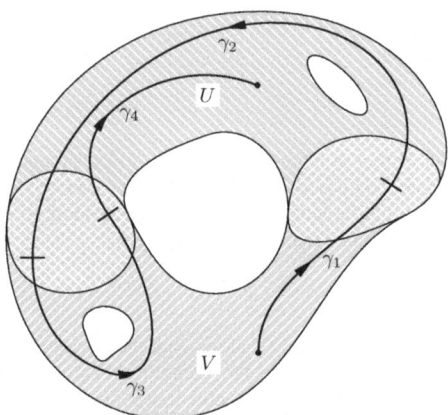

a diagram of solid functors so that the outer square commutes. Then it is to be shown that there is a unique dashed functor $F\colon \Pi(U \cup V) \to \Gamma$ such that the two triangles also commute. For this, we consider finite sequences

$$((\gamma_n, X_n), \ldots, (\gamma_1, X_1)),$$

where X_j is either U or V, and γ_j is a path in X_j. These paths should be composable in $U \cup V$. Therefore, for $k = 1, \ldots, n-1$, the end of γ_k must match the beginning of γ_{k+1}. Then in any case

$$\Pi(j_{X_n})[\gamma_n] \circ \cdots \circ \Pi(j_{X_1})[\gamma_1]$$

is a morphism in $\Pi(U \cup V)$, and the above sequence shall be called a *decomposition* in this proof (Fig. 7.6).

On decomposable morphisms, the F must necessarily be defined by

$$F_{X_n}[\gamma_n] \circ \cdots \circ F_{X_1}[\gamma_1]$$

To show the uniqueness of F, it is now sufficient to show the decomposability of all morphisms of $\Pi(U \cup V)$. Every morphism in $\Pi(U \cup V)$ can be represented by a path γ in $U \cup V$. After a suitable choice of a Lebesgue number (see Sect. 4.1), there is a subdivision of the interval such that the subintervals of γ are mapped entirely to U or entirely to V. A reparametrisation of the subintervals then provides paths in U and V whose composition is a reparametrisation of γ, in particular homotopic with fixed starting and ending points. This is the desired decomposition of $[\gamma]$. For the existence of F, it must now be shown that

$$F_{X_n}[\gamma_n] \circ \cdots \circ F_{X_1}[\gamma_1]$$

7.2 The Seifert–van Kampen Theorem

does not depend on the decomposition of

$$\Pi(j_{X_n})[\gamma_n] \circ \cdots \circ \Pi(j_{X_1})[\gamma_1].$$

There are some rules for how a given sequence can be modified so that it still describes the same morphism and also delivers the same element in Γ. First, notice that, in any decomposition, a pair (γ, U) can be replaced by a pair (γ', U) if $[\gamma] = [\gamma']$ in $\Pi(U)$ holds, and vice versa just as for V instead of U. Furthermore, it is clear that, in a decomposition, a pair (γ, U) may be exchanged by a pair (γ, V) if γ entirely runs in $U \cap V$. This follows immediately from the commutativity of the outer square and vice versa. Finally, in a decomposition, the term $(\ldots, (\gamma, U), (\gamma', U), \ldots)$ can be replaced by $(\ldots, (\gamma\gamma', U), \ldots)$; and vice versa. This follows immediately from the functoriality of F_U; the same for V instead of U. Now let there be two decompositions $(\ldots, (\gamma_k, X_k), \ldots)$ and $(\ldots, (\gamma'_l, X'_l), \ldots)$ of the same morphism of $\Pi(U \cup V)$. It has to be shown that the above construction delivers the same morphism of Γ for both. The three rules above can be used to transform one decomposition into another. First, constant paths can be inserted to ensure that both decompositions have the same number of parts. By composing the γ_k and the γ'_l, two paths γ and γ' are obtained in $U \cup V$ that represent this morphism. There is then a homotopy $H: I \times I \to U \cup V$ from γ to γ' that leaves the start and end point fixed. After a suitable choice of a Lebesgue number, there is a subdivision of the square into smaller rectangles so that H maps each of the rectangles entirely to U or entirely to V. It can be achieved that the subdivision of the path coordinate is the same as the subdivision by the decompositions of the paths γ and γ'. It is now sufficient to show that, by applying the rules from the decomposition of γ, we can successively reach the decomposition of γ' over each rectangle. The first rectangle in Fig. 7.7 provides a bounded homotopy $\gamma_{21} e(\gamma_1(0)) \simeq \gamma_{12}\gamma_1$. For an explicit formula, see the proof of Theorem 7.1.10. This is a relation in $\Pi(U)$ (or $\Pi(V)$) and is therefore respected by F_U (or F_V). Hence, we can swap the two paths and arrive at the next rectangle.

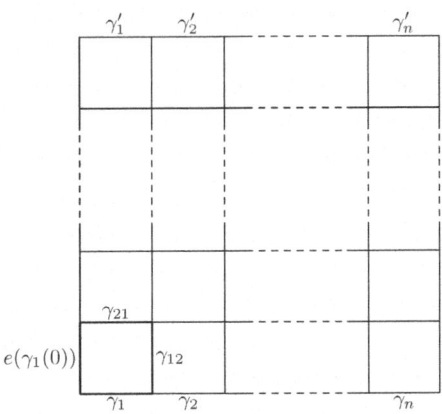

Fig. 7.7 A subdivision of the square into smaller rectangles

We then proceed line by line to reach γ'. To conclude, we must show that the thus defined F is a functor. This follows immediately from the fact that the choice of decomposition is free. □

The theorem allows the following generalisation that will help calculate fundamental groups.

Corollary 7.2.5
For a subset A of X, let $\Pi_A(X)$ be the full subgroupoid with objects in A (and the same sets of morphisms). Suppose A meets every path component of U, V and $U \cap V$. Then

$$\begin{array}{ccc} \Pi_A(U \cap V) & \xrightarrow{\Pi(i_V)} & \Pi_A(V) \\ \Pi(i_U) \downarrow & & \downarrow \Pi(j_V) \\ \Pi_A(U) & \xrightarrow{\Pi(j_U)} & \Pi_A(U \cup V) \end{array}$$

is a pushout.

Proof. First, we construct left inverses $R_{U \cap V}$, R_U, R_V and $R_{U \cup V}$ to the respective inclusion functors of the subgroupoids as follows: for each x from $U \cap V$, choose a path in $U \cap V$ to a point in A that should be called $R_{U \cap V}(x)$. For $x \in A$, this path should be constant. Then for each x from $U \setminus V$, choose a path in U to $R_U(x) \in A$. Similarly for V. The chosen paths to the points x from $U \cup V$ are denoted by Φ_x. This also determines the retraction functors on the morphisms (Fig. 7.8).

We have
$$R_{U \cap V} \circ I_{U \cap V} = \mathrm{id}_{\Pi_A(U \cap V)}$$

Fig. 7.8 Left inverse retraction functors

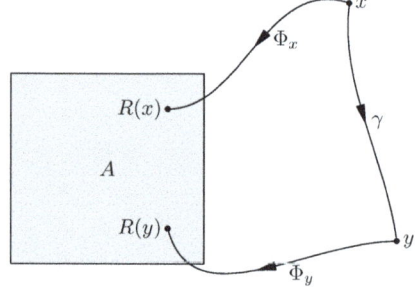

7.2 The Seifert–van Kampen Theorem

and correspondingly for the other open sets. Therefore, the diagram induced by the inclusions for Π_A is a retract of the diagram for Π. The assertion now follows from the general observation that retracts of pushouts are always pushouts. To see this, suppose the inner square of the diagram

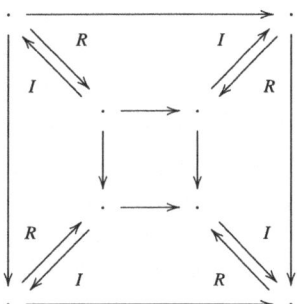

is a retract of the outer one. To construct an arrow from the object at the bottom right of the inner rectangle to another object, we can transfer the given arrows at the other corners to the outer rectangle with the help of the retraction morphism R. Now, we use the universal property and compose the obtained arrow with the arrow I. The relation $RI = \mathrm{id}$ then shows the commutativity and the uniqueness. □

Example 7.2.6
Cover the circle $S^1 \subset \mathbb{C}$ by the two open subsets $U = S^1 \setminus \{-i\}$ and $V = S^1 \setminus \{i\}$. Then $A = \{-1, 1\}$ meets all path components of the covering sets and the intersection. Since U is contractible the groupoid $\Pi_A(U)$ consists of one isomorphism α_1 between 1 and -1, its inverse and the identities. Similarly, the groupoid $\Pi_A(V)$ consists of an isomorphism α_2 between -1 and 1, its inverse and the identities. The groupoid of the intersection has only identities. Define the groupoid $\mathcal{G}$ with object set A and morphisms

$$\mathrm{Mor}_{\mathcal{G}}(1, 1) = \{\beta^n \mid n \in \mathbb{Z}\}$$
$$\mathrm{Mor}_{\mathcal{G}}(1, -1) = \{\alpha_1 \beta^n \mid n \in \mathbb{Z}\}$$
$$\mathrm{Mor}_{\mathcal{G}}(-1, 1) = \{\beta^n \alpha_1^{-1} \mid n \in \mathbb{Z}\}$$
$$\mathrm{Mor}_{\mathcal{G}}(-1, -1) = \{\alpha_1 \beta^n \alpha_1^{-1} \mid n \in \mathbb{Z}\},$$

where $\beta = \alpha_2 \alpha_1$ is represented by the path e_1 that runs once in the mathematically positive sense around the circle. In the diagram

$$\begin{array}{ccc} \Pi_A(U \cap V) & \longrightarrow & \Pi_A(U) \\ \downarrow & & \downarrow \\ \Pi_A(V) & \longrightarrow & \mathcal{G} \end{array}$$

the lower left functor is given by $\alpha_2 \mapsto \beta \alpha_1^{-1}$ and the right vertical functor by $\alpha_1 \mapsto \alpha_1$. This diagram is a pushout: to construct a functor F from $\mathcal{G}$ from given F_U and F_V, it is sufficient to specify its values on α_1 and on β. But these are determined as follows:

$$F(\alpha_1) = F_U(\alpha_1)$$
$$F(\beta) = F_V(\alpha_2)F_U(\alpha_1).$$

So we have

$$\Pi_A(S^1) \cong \mathcal{G},$$

and we obtain another proof for

$$\pi_1(S^1, 1) \cong \{\beta^n \mid n \in \mathbb{Z}\} \cong \mathbb{Z}.$$

Corollary 7.2.7 (Seifert–van Kampen for Fundamental Groups)
Let X be a topological space that is covered by open sets U and V. Let x be a point in $U \cap V$, and let $U \cap V$ be path-connected. Then the diagram

$$\begin{array}{ccc} \pi_1(U \cap V, x) & \xrightarrow{\pi_1(i_V)} & \pi_1(V, x) \\ {\scriptstyle \pi_1(i_U)}\downarrow & & \downarrow {\scriptstyle \pi_1(j_V)} \\ \pi_1(U, x) & \xrightarrow[\pi_1(j_U)]{} & \pi_1(U \cup V, x) \end{array}$$

is a pushout (of groups).

Proof. Let X' be the path component of x in X. Set $U' = U \cap X'$ and $V' = V \cap X'$. We want to apply the previous conclusion to $X' = U' \cup V'$ and $A = \{x\}$. It should be noted that A hits all path components: for $y \in U'$, there is a path in X' that connects y with x. However, this could take values outside of U'. If this is the case, take only a beginning piece of the path to get to $U' \cap V'$. Since $U' \cap V'$ is path-connected the path can be extended to x without falling into the complement of U'. Similarly, we can see that A also hits all path components of V' (Fig. 7.9). □

Fig. 7.9 For the proof of Seifert–van Kampen for fundamental groups

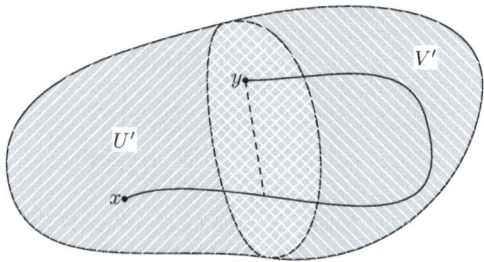

7.2 The Seifert–van Kampen Theorem

The preceding conclusion implies that X is simply-connected if there is a cover by open, simply-connected subsets U and V such that $U \cap V$ is path-connected. (It is not necessary for $U \cap V$ to be simply-connected; the trivial group is a pushout also in the more general situation.) From the obvious cover of the spheres by the complements of the North and South Pole, we get:

Theorem 7.2.8
For all $n \geq 2$, the sphere S^n is simply-connected.

The result about spheres can now be used to study the behaviour of the fundamental group when attaching cells to a space X. Let X be path-connected and $f \colon S^{n-1} \to X$ a continuous map for a $n \geq 1$. Then we consider a pushout

$$\begin{array}{ccc} S^{n-1} & \xrightarrow{\subseteq} & D^n \\ f \downarrow & & \downarrow \\ X & \longrightarrow & Y. \end{array}$$

To cover the space Y with open sets, proceed as follows: let V be the interior of the n-cell. Let the set U be the complement of the centre of the n-cell

$$U = X +_f D^n \setminus 0.$$

Then V is contractible, and U is homotopy equivalent to X (Fig. 7.10).

Fig. 7.10 A retraction of U to X

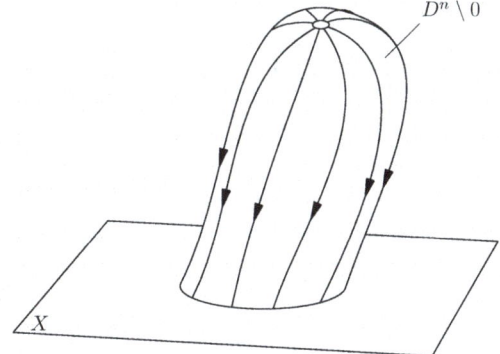

The intersection is homotopy equivalent to S^{n-1}. The theorem of Seifert–van Kampen (in the version for fundamental groups) can, therefore, only be applied if $n \geqslant 2$. Then we obtain a pushout

$$\begin{array}{ccc} \pi_1(S^{n-1}, p) & \longrightarrow & 1 \\ {\scriptstyle f_*}\downarrow & & \downarrow \\ \pi_1(X, f(p)) & \longrightarrow & \pi_1(Y, f(p)) \end{array}$$

in the category of groups. (Here, the point p is any point of the sphere.) If $n \geqslant 3$, then $\pi_1(S^{n-1}, p)$ is also trivial; the inclusion $X \to Y$ therefore induces an isomorphism of fundamental groups. If $n = 2$, then $\pi_1(S^1, p) \cong \mathbb{Z}$, and the induced homomorphism $f_* : \pi_1(S^1, p) \to \pi_1(X, f(p))$ maps the generator to the homotopy class of the loop described by f. In the pushout Y, this is null-homotopic. This follows from the commutativity of the diagram but should also be intuitively clear: if we attach a 2-cell using f, the map f can be null-homotoped over this 2-cell. It follows from the universal property of the pushout that $\pi_1(Y, f(p))$ is isomorphic to the quotient group of $\pi_1(X, f(p))$ by the normal subgroup generated by this class:

$$\pi_1(Y, f(p)) \cong \pi_1(X, f(p))/N(f_*\pi_1(S^1, p)) = \pi_1(X, f(p))/N(\langle f_*e_1\rangle).$$

Not every group element generates a normal subgroup. Simple examples can be found in the symmetric groups. Given a transposition, we obtain every other transposition through conjugation, and these generate the whole symmetric group. The normal subgroup generated by a transposition is, therefore, the entire group, while the subgroup it generates only has two elements.

Supplement

Realisations of Fundamental Groups It is now easy to construct spaces whose fundamental groups are isomorphic to $\mathbb{Z}/k$ for $k \geqslant 2$; we simply attach a 2-cell to S^1 with a map of degree k, such as e_k. By the way, these spaces are called n-fold *fool's caps*. It is known that every finitely generated abelian group is isomorphic to a product of cyclic groups. Using products of fool's caps, we can realise any such group up to isomorphism as a fundamental group. In fact, any group can be realised as the fundamental group of a space, and not just ad hoc but functorially (see Sect. 11.3).

Exercises

Exercise 118 Free Products
Show that the free product of two non-trivial groups is never abelian and never finite.

7.3 Surfaces 147

Exercise 119 An Eight
Let x, y be two points on the circle. The one-point union of two circles in x, y is the quotient space

$$(S^1, x) \vee (S^1, y) = (S^1 + S^1)/x \sim y.$$

Calculate the fundamental group.

Exercise 120 Mayer and Vietoris Never Give Up
If, in the theorem of Seifert–van Kampen, it is additionally assumed that all occurring fundamental groups are abelian, then we arrive at an exact sequence

$$\pi_1(U \cap V, x) \longrightarrow \pi_1(U, x) \oplus \pi_1(V, x) \longrightarrow \pi_1(U \cup V, x) \longrightarrow 0.$$

(See Sect. 9.3.5 for the concept of an exact sequence.) Construct this sequence and give an example where the homomorphism on the left is not injective.

7.3 Surfaces

This section aims to apply the previously presented theory to some geometrically relevant examples.

Unions with 1-Cells Before we get to the surfaces, we first consider the situation where the components of a sum $X = X_1 + X_2$ of path-connected spaces are connected by attaching a 1-cell. The pushout is covered by two open sets U and V that consist of either X_1 or X_2 and the interior of the 1-cell. The intersection is the interior of the interval, thus contractible. The fundamental group of the resulting space is therefore the free product

$$\pi_1(X_1, x_1) * \pi_1(X_2, x_2)$$

of the fundamental groups of X_1 and X_2 (see Sect. 7.2). For example, if X_1 and X_2 are both homeomorphic to the circle S^1, we obtain the *free group* $\mathbb{Z} * \mathbb{Z}$ on two generators. If a and b are the generators, this group consists of all words in the symbols 1, a, a^{-1}, b and b^{-1} with the obvious reduction rules. A homomorphism from $\mathbb{Z} * \mathbb{Z}$ to another group G is then equivalent to a pair of elements from G, namely the two images of the generators. Since there are groups that are not abelian, we have $ba \neq ab$ in $\mathbb{Z} * \mathbb{Z}$. This group is, therefore, not abelian itself. Inductively, we can also construct the free group on n generators. (The space used in the process is, by the way, homotopy equivalent to the complement of n points in $\mathbb{R}^2$.) More generally, it is not hard to see that the fundamental groups of (realisations of) quivers are always free.

The Torus For the torus $S^1 \times S^1$ we obtain

$$\pi_1(S^1 \times S^1) \cong \mathbb{Z} \oplus \mathbb{Z}$$

Fig. 7.11 To the attaching map of the torus

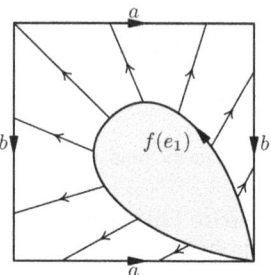

from the compatibility of π_1 with products. The class corresponding to (m, n) is represented by the loop $z \mapsto (z^m, z^n)$. We can determine the fundamental group of the torus in another way. To do this, we note that the torus contains the subspace

$$S^1 \times \{1\} \cup \{1\} \times S^1.$$

Its fundamental group is $\mathbb{Z} * \mathbb{Z}$. This follows from the fact that the space is homotopy equivalent to the space created by connecting two circles by a 1-cell. We can also see it directly with an open cover (see exercise 'An eight' in Sect. 119). The complement of this subspace is

$$S^1 \setminus 1 \times S^1 \setminus 1,$$

hence a 2-cell. Hence, the torus is created by attaching a 2-cell to $S^1 \times \{1\} \cup \{1\} \times S^1$ and we conclude that its fundamental group is isomorphic to the quotient of $\mathbb{Z} * \mathbb{Z}$ by the relation given by the attaching map. The attaching map is easy to determine: if a and b are the generators of $\mathbb{Z} * \mathbb{Z}$ then the boundary of the 2-cell is attached along the commutator

$$[a, b] = aba^{-1}b^{-1},$$

as shown in Fig. 7.11.

Thus, the group $\pi_1(S^1 \times S^1)$ is the quotient group of the free group on two generators a, b by the normal subgroup generated by $[a, b]$. We denote this simply by

$$\langle a, b \mid [a, b] \rangle.$$

This notation lists the generators first, followed by the relation $[a, b] = 1$. As noted above, this group is isomorphic to the free abelian group in two generators.

7.3 Surfaces

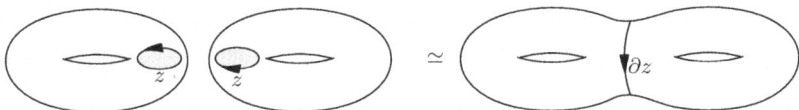

Fig. 7.12 The pretzel surface as a connected sum of two tori

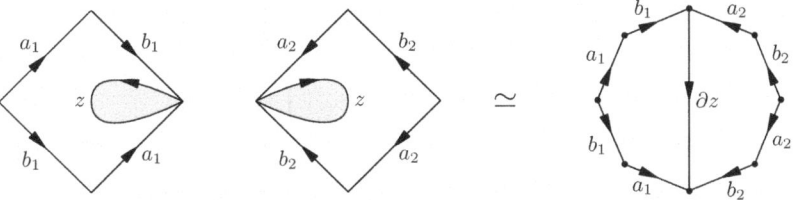

Fig. 7.13 The fundamental group of the pretzel surface

Connected Sums The second approach to the fundamental group of the 2-torus has the advantage that the following remark is evident. If

$$z \colon D^2 \to S^1 \times S^1$$

is an embedding, then the restriction to the boundary represents an element of the fundamental group of $(S^1 \times S^1) \setminus z(0)$ and this is just the commutator of the generators—or its inverse, depending on the orientation. If we next take two tori with embedded disks and connect them by surgery (see Sect. 2.4) as in Fig. 7.12 we speak of the *connected sum*. It is called the *pretzel surface* and is denoted by

$$(S^1 \times S^1) \, \# \, (S^1 \times S^1).$$

The fundamental group of the pretzel surface can be determined using the theorem of Seifert–van Kampen. To do this, we again cover the surface with an open disk and the complement U of a point in the disk. Since U is homotopy equivalent to a sum of four circles united at one point, its fundamental group is free on generators a_1, b_1 and a_2, b_2. The intersection is homotopy equivalent to a circle, and this circle corresponds in U to the product of the commutators $[a_1, b_1][a_2, b_2]$ (Fig. 7.13).

The fundamental group of the pretzel surface is thus described by

$$\langle a_1, b_1, a_2, b_2 \mid [a_1, b_1][a_2, b_2] \rangle.$$

The preceding surgery can be iterated. This yields a surface

$$\underbrace{(S^1 \times S^1) \# \ldots \# (S^1 \times S^1)}_{g}.$$

It depends—strictly speaking—on the choice of the embedded discs. However, up to homeomorphism, the choice of embeddings does not matter. Therefore, we allow ourselves to use the notation F_g for this surface. The number g is called the *genus* of the surface; it denotes the number of tori used. Therefore, the space F_0 is a sphere and F_1 is a torus. In fact, every connected, orientable, closed surface is homeomorphic to a surface F_g. However, this will not be shown here. We content ourselves with recording the isomorphism type of the fundamental group

$$\pi_1(F_g) \cong \langle a_1, b_1, \ldots, a_g, b_g \mid \prod_{j=1}^{g} [a_j, b_j] \rangle.$$

This follows inductively as above in the case $g = 2$. These groups are not easy to understand. They can be simplified by making them abelian, i.e., by factoring out the normal subgroup generated by all commutators. The above relation then becomes trivial, and we obtain the free abelian group on $2g$ generators:

$$\pi_1^{\mathrm{ab}}(F_g) = \pi_1(F_g)/[\pi_1(F_g), \pi_1(F_g)] \cong \mathbb{Z}^{\oplus 2g}.$$

From $\pi_1(F_g) \cong \pi_1(F_h)$ follows $\pi_1^{\mathrm{ab}}(F_g) \cong \pi_1^{\mathrm{ab}}(F_h)$. If F_g and F_h are homotopy equivalent, then $g = h$ must hold. If, on the contrary, we have $g \neq h$, then F_g is not homotopy equivalent to F_h and certainly not homeomorphic to it.

Non-orientable Surfaces Next, we consider the projective plane $\mathbb{R}P^2$. The disk D^2 is homeomorphically mapped onto the upper hemisphere in S^2. If we identify opposite points on the edge of the disk, then the usual argument (once again the corollary of Sect. 4.1 shows that this space is homomorphic to $\mathbb{R}P^2$ (Fig. 7.14). The now repeatedly applied procedure yields

$$\pi_1(\mathbb{R}P^2) \cong \langle a \mid a^2 \rangle \cong \mathbb{Z}/2.$$

Similarly, we can proceed with the connected sum of several projective planes. If N_k denotes the k-fold connected sum, then we get

$$\pi_1(N_k) \cong \langle a_1, a_2, \ldots, a_k \mid a_1^2 a_2^2 \cdots a_k^2 \rangle.$$

The case $k = 2$ is outlined in Fig. 7.15.

Fig. 7.14 The projective plane as a quotient of the disk

7.3 Surfaces

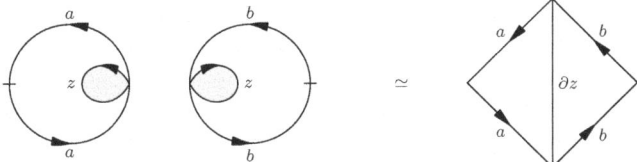

Fig. 7.15 $\pi_1(N_2) \cong \langle a, b \mid a^2 b^2 \rangle$

None of the groups $\pi_1(N_k)$ is isomorphic to a fundamental group of a surface F_g because even after abelianisation, the product of the generators has order 2. Such an element does not exist in a free abelian group. To show that they differ, we consider the groups $\mathrm{Hom}(\pi_1(N_k), \mathbb{Z}/2)$. We have

$$\mathrm{Hom}(\pi_1(N_k), \mathbb{Z}/2) \cong \mathrm{Hom}(\pi_1^{\mathrm{ab}}(N_k), \mathbb{Z}/2) \cong \mathrm{Hom}(\mathbb{Z}^k, \mathbb{Z}/2) \cong (\mathbb{Z}/2)^k.$$

In the middle isomorphism, we used that the homomorphisms f to $\mathbb{Z}/2$ always satisfy the relation

$$f(a_1^2 a_2^2 \cdots a_k^2) = 2(f(a_1) + f(a_2) + \cdots + f(a_k)) = 0.$$

Thus, the surfaces N_k cannot be homotopy equivalent amongst each other.

Supplements

The Automorphism Group of the Torus The group $\mathrm{Aut}(S^1 \times S^1)$ acts transitively on the torus. Let (g, h) be a point. Then the homomorphism

$$S^1 \times S^1 \longrightarrow S^1 \times S^1, \quad (x, y) \longmapsto (gx, hy)$$

maps the point $(1, 1)$ to (g, h). If H denotes the stabiliser of $(1, 1)$ then group multiplication provides a homeomorphism

$$\mathrm{Aut}(S^1 \times S^1) \cong (S^1 \times S^1) \times H.$$

Every element of H maps the point $(1, 1)$ to itself, thus inducing an automorphism of the fundamental group $\pi_1(S^1 \times S^1) \cong \mathbb{Z} \oplus \mathbb{Z}$. This results in a (continuous) homomorphism from H to $\mathrm{GL}(2, \mathbb{Z})$. Every matrix

$$\begin{pmatrix} a & b \\ c & d \end{pmatrix}$$

Fig. 7.16 A (non-pure) braid

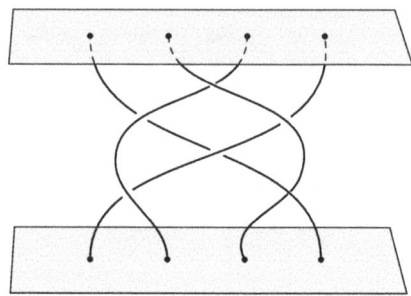

from $GL(2, \mathbb{Z})$ provides a homeomorphism

$$S^1 \times S^1 \longrightarrow S^1 \times S^1, \ (x, y) \longmapsto (x^a y^b, x^c y^d)$$

of $S^1 \times S^1$ that we can describe by this matrix on the fundamental group. Overall, this results in a homeomorphism

$$\text{Aut}(S^1 \times S^1) \cong GL(2, \mathbb{Z}) \times (S^1 \times S^1) \times H',$$

where H' is the subgroup of homeomorphisms that fix $(1, 1)$ and induce the identity on the fundamental group. It can be shown that H' is path-connected (even contractible). The mapping class group of the torus is therefore isomorphic to $GL(2, \mathbb{Z})$. The determinant of a matrix from $GL(2, \mathbb{Z})$ is ± 1. The homeomorphisms with positive determinants are orientation-preserving; the others are not.

Braid Groups Let F be a surface, and S be a finite subset. Let $\text{Emb}(S, F) \subset \text{Hom}(S, F)$ denote the space of embeddings, i.e., the injective maps from S to F. Its fundamental group $\pi_1(\text{Emb}(S, F), S)$ is the *pure braid group* of F on n strands. Braids can be well visualised for $F = \mathbb{R}^2$ (see Fig. 7.16).

Exercises

Exercise 121 Nail Test
Someone wants to attach a valuable painting to a wall by hammering two nails into the wall over which a string runs, holding the picture. They hope that if one nail falls out of the wall, the other will still keep the picture in place, albeit somewhat crooked. However, something can go wrong: show that the string can be wrapped around both nails so that the picture falls with each nail.

Exercise 122 The Möbius Strip
A model of the Möbius strip is given by

$$M = \{([x, y], (x', y')) \in \mathbb{RP}^1 \times D^2 \mid x'y = xy'\}.$$

7.3 Surfaces

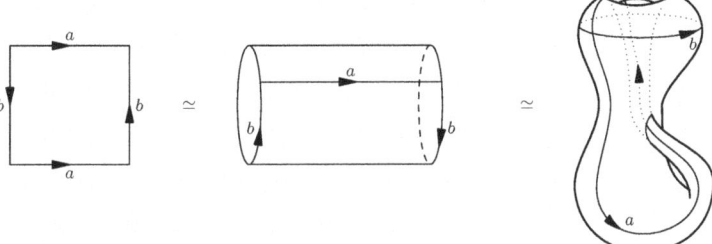

Fig. 7.17 The Klein bottle

How can the fundamental group of M be determined? Which element is described by the boundary curve

$$S^1 \longrightarrow M, \ (x, y) \longmapsto ([x, y], (x, y))?$$

Show: if we remove a point from the projective plane $\mathbb{R}P^2$ the rest becomes homeomorphic to

$$M' = \{([x, y], (x', y')) \in \mathbb{R}P^1 \times \mathbb{R}^2 \mid x'y = xy'\}.$$

The projective plane is thus created by attaching a 2-cell to the Möbius strip along its boundary.

Exercise 123 And So On
Show that the projective space $\mathbb{R}P^n$ is created by attaching an n-cell to $\mathbb{R}P^{n-1}$ also for the case $n \geqslant 3$. The inclusions, therefore, induce isomorphisms

$$\pi_1(\mathbb{R}P^2) \cong \pi_1(\mathbb{R}P^3) \cong \pi_1(\mathbb{R}P^4) \cong \ldots \cong \pi_1(\mathbb{R}P^\infty).$$

Exercise 124 The Klein Bottle
The Klein bottle K is obtained by identifying the boundary of a square according to the scheme indicated in Fig. 7.17.
Show

$$\pi_1(K) \cong \langle a, b \mid aba^{-1}b \rangle.$$

Also, show that K is homeomorphic to N_2 by cutting the square along a diagonal and glueing the two parts back together in b.

Exercise 125 Connected Sum
Let X be covered by two open sets U and V such that $U \cap V$ is homeomorphic to $S^1 \times \mathbb{R}$. What does the Seifert–van Kampen theorem say for this situation? Use this to (once again) determine the fundamental group of the pretzel surface.

Exercise 126 Heegaard Cooks Lentils
Let L be the pushout

$$\begin{array}{ccc} S^1 \times S^1 & \xrightarrow{\subseteq} & D^2 \times S^1 \\ {\scriptstyle f}\downarrow & & \downarrow \\ S^1 \times D^2 & \longrightarrow & L, \end{array}$$

where $f(z, w) = (z^a w^b, z^c w^d)$ with integers a, b, c, d and $ad - bc = \pm 1$. Then the fundamental group of L is isomorphic to $\mathbb{Z}/a$. (The space L will return as $L(a; p, q)$ for suitable p and q in Sect. 9.3.) Such a diagram is called a *Heegaard splitting* of L, after the mathematician who showed in his dissertation that all three-manifolds can be similarly decomposed.

Covering Spaces

8

When determining the fundamental group of the circle in Sect. 6.4, we considered the exponential map, which laid out the real numbers like a helix over the circle and thus 'covered' it. The maps we will consider in this chapter are generalisations of this situation. The lifting behaviour of paths in coverings can be used to calculate fundamental groups. The connection between the fundamental group and coverings is even closer and leads to the classification of coverings in terms of the fundamental group. The whole theory is analogous to the Galois theory of field extensions.

8.1 The Category of Coverings

Definition 8.1.1

A *covering* of a space B is a continuous map $p \colon X \to B$ with the following properties:

(Cov 1) Discrete: for each $b \in B$, the fibre $p^{-1}(b)$ is discrete.
(Cov 2) Locally trivial: For each $b \in B$ there is a neighbourhood U and a homeomorphism h_U from $p^{-1}(U)$ to $U \times p^{-1}(b)$ that makes the diagram

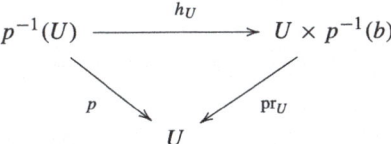

commutative.

The pre-image $p^{-1}(b)$ is called the *fibre* over b and the map h_U is called a *trivialisation* above U. Because the fibre is discrete, the space $U \times p^{-1}(b)$ is

Fig. 8.1 The sheets of a covering

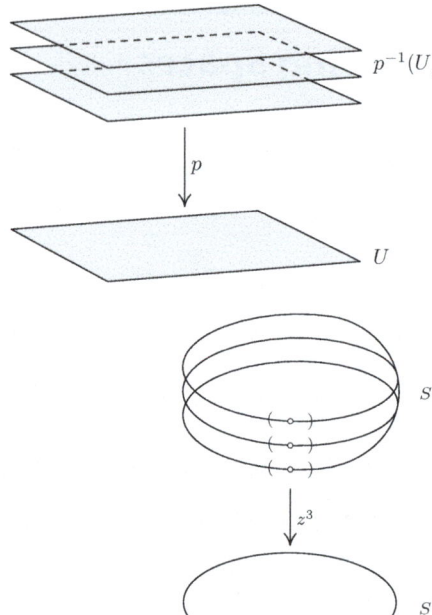

Fig. 8.2 The covering $z \mapsto z^3$ of the circle S^1

a topological sum; there is a summand that looks like U for each point of the fibre. The summands are also called *sheets* of the covering (see Fig. 8.1).

The local triviality implies the following result.

▶ **Remark 8.1.2** A covering is a local homeomorphism.

Examples 8.1.3
The simplest examples of coverings of B are homeomorphisms $B' \cong B$ and, more generally, the projections $\mathrm{pr}_B : B \times F \to B$ with discrete spaces F. It should be noted that a covering does not have to be surjective, as the example $\emptyset \to B$ shows. More interesting coverings are the maps

$$S^1 \longrightarrow S^1, \ z \longmapsto z^n$$

for $n \neq 0$. Each fibre consists of n elements. As trivialising neighbourhoods, all open circular arcs of length less than $2\pi/n$ can be used (see Fig. 8.2).
The exponential map

$$\mathbb{R} \longrightarrow S^1, \ t \longmapsto \exp(2\pi i t)$$

is a covering with countable fibres. Here, we can even take open circular arcs of length less than 2π. The identification $S^n \to \mathbb{R}P^n$ is a two-sheeted covering, because if V is an open subset of S^n without antipodal points, then for $U = p(V)$

$$p^{-1}U = V + (-V) \cong U \times \mathbb{Z}/2.$$

8.1 The Category of Coverings

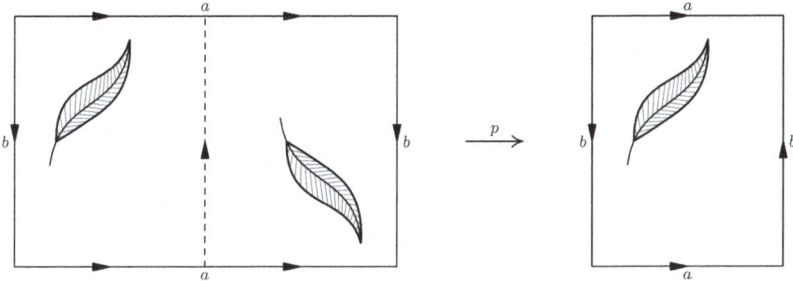

Fig. 8.3 A two-sheeted covering of the Klein bottle by a torus

Figure 8.3 shows a two-sheeted covering of the Klein bottle by the torus.
These examples are all locally trivial but not globally because the covering spaces are connected, and the fibres consist of several elements.

There are many ways to construct new coverings from old ones, for example:

▶ **Remark 8.1.4** Every base change of a covering is a covering.

This is true because base change preserves the property of being a product with a discrete space.

The composition of coverings does not need to be a covering. However, under an additional condition, this is correct.

Theorem 8.1.5
If $p: X \to B$ and $f: B \to C$ are coverings, and if f has finite fibres, then the composition $fp: X \to C$ is also a covering.

Proof. Let c be a point of C and V an open neighbourhood over which f is trivial. A trivialisation $f^{-1}(V) \cong V \times f^{-1}(c)$ then provides for each point b of the fibre $f^{-1}(c)$ an open neighbourhood U_b of b in B that is mapped homeomorphically onto V under f. Each U_b then contains an open neighbourhood U'_b of b, such that p is trivial above U'_b. Then

$$V' = \bigcap_{b \in f^{-1}(c)} f(U'_b)$$

is an open neighbourhood of c in C that is contained in V. Consequently, the map f is trivial over V' and p is trivial over $f^{-1}(V')$. Thus, the map fp is trivial over V. □

So, while coverings can be pulled back along any continuous map, they can only be pushed forward along finite coverings.

Definition 8.1.6

The map fp is also called the *transfer* of p along f.

It should be noted that the cardinality of the fibres can change during the transfer.

▶ **Remark 8.1.7** If $p\colon X \to B$ and $p'\colon X' \to B'$ are coverings, then

$$p + p'\colon X + X' \longrightarrow B + B'$$

and

$$p \times p'\colon X \times X' \longrightarrow B \times B'$$

are also coverings.

Definition 8.1.8

The constructions of the preceding note are also called *external sum* and *external product*, because the base changes accordingly. The *internal product* of two coverings $p\colon X \to B$ and $q\colon Y \to B$ over the same base are obtained from the external product by base change along the diagonal

$$\Delta = (\mathrm{id}, \mathrm{id})\colon B \longrightarrow B \times B.$$

The fibre in the internal product over a point b of B is then the product of the fibres of the factors over b. The *internal sum* of two coverings $p\colon X \longrightarrow B$ and $q\colon Y \to B$ over the same base are obtained from the external sum by transfer along the co-diagonal

$$\nabla = (\mathrm{id}, \mathrm{id})\colon B + B \longrightarrow B.$$

The fibre of the internal sum over a point b of B is then the sum of the fibres of the factors over b.

Example 8.1.9

Consider the external product of the homeomorphism

$$\mathbb{R} \longrightarrow {]0, \infty[}, \quad x \longmapsto \exp(x)$$

with the covering

$$\mathbb{R} \longrightarrow S^1, \quad y \longmapsto \exp(iy).$$

8.1 The Category of Coverings

Fig. 8.4 Branches of the logarithm

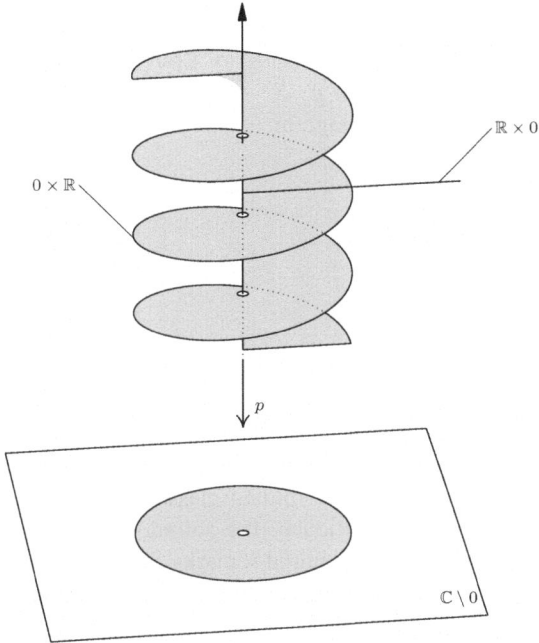

This agrees after the composition with the homeomorphism

$$]0, \infty[\times S^1 \longrightarrow \mathbb{C} \setminus 0, \ (t, z) \mapsto tz$$

with the complex exponential function

$$\mathbb{R} \times \mathbb{R} = \mathbb{C} \longrightarrow \mathbb{C} \setminus 0, z = x + iy \mapsto \exp(z) = \exp(x) \exp(iy).$$

In particular, the complex exponential function exp is a covering. (Figure 8.4 shows how the imaginary axis $\{0\} \times \mathbb{R}$ unwinds over the unit circle, while the real axis $\mathbb{R} \times \{0\}$ is mapped as usual.) So, locally, we can always choose a sheet that maps homeomorphically onto an open subset of the punctured plane. A local inverse function of this map is also called a *branch of the logarithm*. More about coverings in function theory can be found in the supplement.

The totality of all coverings of a topological space B is best organised in a category.

Definition 8.1.10

The objects of the category **Cov**(B) are the coverings $p\colon X \to B$ of B. The morphisms from $p\colon X \to B$ to $q\colon Y \to B$ are the continuous maps $f\colon X \to Y$ that make the diagram

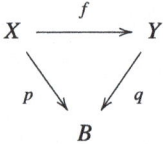

commutative.

The classification problem for coverings now is the problem of describing the category of coverings for a given topological space B. To get an overview of the possible isomorphism classes, the morphisms between coverings must be understood. In particular, the automorphism groups are again interesting. To this end, we make some initial remarks.

Definition 8.1.11

If $p\colon X \to B$ is a covering, then the automorphism group of p in the category of coverings of B is a subgroup of the homeomorphism group of X. It is called the *deck transformation group* of p and is endowed with the discrete topology.

The deck transformation group acts continuously on X. The fibres of p are invariant. Thus, we obtain a continuous map

$$X/\mathrm{Aut}(p) \longrightarrow B$$

However, this map is only a homeomorphism in certain cases, as we will see in Theorem 8.5.7.

Supplement

Riemann Surfaces In function theory, 'multi-valued functions' often occur, such as the n-root functions or the logarithm discussed earlier. These constructions are made 'single-valued', i.e., they are turned into actual functions by passing to covering spaces of the domain. We will briefly outline this process below. For simplicity's sake, we assume that the 'multi-valued function' is implicitly defined by a polynomial equation $r(x, y) = 0$ in two complex variables. In the case of the root function $y = \sqrt{x}$, this is for example $y^2 = x$, thus $r(x, y) = x - y^2$. Another interesting example is $r(x, y) = x^3 - x - y^2$. If $y = f(x)$ for a 'single-valued' polynomial function f, then of course $r(x, y) = y - f(x)$. Normally, the equation $r(x, y) = 0$ cannot be solved for y, so that we do not have a function $\mathbb{C} \to \mathbb{C}$. Instead, we consider the solution set R of the equation $r(x, y) = 0$ in $\mathbb{C}^2$ along with its two projections to $\mathbb{C}$:

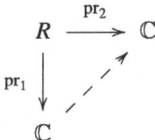

If r is given by a function f, then R is its graph, and we have $f\mathrm{pr}_1 = \mathrm{pr}_2$, which is indicated by the dashed arrow. In general, there is no such f, and we will have to interpret the diagram in such a way that pr_1 gives an extension of the domain so that we can use pr_2 to define a function on R. We then call R the *Riemann surface* for the 'multi-valued function' given by r. Of course, we must make further assumptions on r to ensure that it is actually a smooth surface (2-dimensional manifold). In the points (x, y) of R, the derivatives $\partial r/\partial x$ and $\partial r/\partial y$ may not both vanish. In the points with $\partial r/\partial y \neq 0$, the map pr_1 is a local homeomorphism. If we only consider these points of R, the restriction of pr_1 provides a covering of its image. In general, we have to speak of *branched coverings*. These are not coverings in the sense we defined but generalisations thereof.

Exercises

Exercise 127 When the Negative Is Missing
The map
$$]0, \infty[\longrightarrow S^1, \quad t \longmapsto \exp(2\pi i t)$$
is not a covering.

Exercise 128 Regular
Let X and Y be compact Hausdorff spaces, and let $f \colon X \to Y$ be a local homeomorphism. Then f is a covering with finite fibres.

Exercise 129 Irregular
Provide an example of a surjective local homeomorphism that is not a covering.

Exercise 130 Untangle
Describe explicitly a double covering of the Möbius strip by the unwound band $S^1 \times I$.

Exercise 131 Repetition
Let $p \colon X \to Y$ be a covering. Prove or disprove:

(1) If Y is compact, then so is X.
(2) If X is compact, then so is Y.

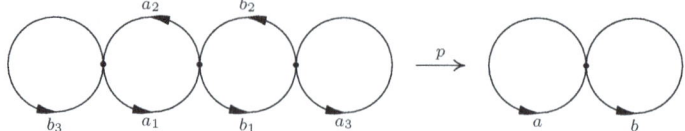

Fig. 8.5 A three-sheeted covering of $S^1 \vee S^1$

(3) If Y is a Hausdorff space, then so is X.
(4) If X is a Hausdorff space, then so is Y.

Exercise 132 Complex, But Not Complex
Is there a polynomial function $p \colon \mathbb{C} \to \mathbb{C}$ with $\deg(p) \geqslant 2$ that is a covering?

Exercise 133 Roundabout
For $n \neq 0$ let $e_n \colon S^1 \to S^1$ be the covering given by $z \mapsto z^n$. Then e_n and e_{-n} are isomorphic coverings of S^1. How many elements does the group $\operatorname{Aut}(e_n)$ of deck transformations of e_n have? What is its structure?

Exercise 134 Even More Circles
Figure 8.5 describes a three-sheeted covering of the one-point union of two circles $S^1 \vee S^1$. Describe the deck transformation group.

8.2 The Lifting Theorem

In Sect. 6.4, we have already considered paths and loops in the covering $p \colon \mathbb{R} \to S^1$. That discussion was based on the lifting theorem, which is also available for general coverings and is not proven much differently. The uniqueness statement is proven there using the fact that the exponential map is a group homomorphism; the uniqueness problem must be solved differently for general coverings.

> **Theorem 8.2.1**
>
> Let $p \colon X \to B$ be a covering and Z a connected topological space. Then any two maps $F, F' \colon Z \to X$ with $pF = pF'$ agree as soon as they agree in one point.

Proof. If x and y are two points in X with $p(x) = p(y)$, they are in different sheets under any trivialisation. So, they are separated by disjoint neighbourhoods. This implies that the diagonal in $X \times_B X$ is closed. (Coverings are separated, see Exercise 58 in Sect. 3.2 for this notion.) So, if F and F' are continuous maps to X whose compositions with p coincide, then (F, F') is a continuous map to $X \times_B X$. The pre-image of the diagonal is the set of points at which F and F' coincide. This set is thus closed. According to the assumption, it is not empty. Because of the

8.2 The Lifting Theorem

connectivity of Z, it is now sufficient to show that it is open. Let z be any point with $F(z) = F'(z) = x$. Then there is an open neighbourhood V of $p(x)$ in B over which p is trivial. Let

$$h \colon p^{-1}(V) \to V \times p^{-1}(p(x))$$

be a trivialisation. Then because of the continuity of F and F', there is an open neighbourhood U of z with $hF(U) \subseteq V \times \{x\}$ and $hF'(U) \subseteq V \times \{x\}$. Except for the homeomorphism $V \times \{x\} \cong V$, the maps hF and hF' coincide on U with f. Since h is a homeomorphism, the maps F and F' coincide on U. □

Theorem 8.2.2 (Lifting Theorem)
Let $p \colon X \to B$ be a covering and Z a connected topological space. Let $f \colon Z \to X$ be a continuous map and $H \colon I \times Z \to B$ a homotopy with start pf. Then there is a unique homotopy $F \colon I \times Z \to X$ with start f and $pF = H$.

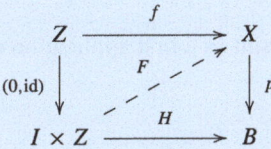

Proof. The uniqueness immediately follows the previous result. It remains to show the existence. First, we consider the case that Z is a point x in X. Then H is a path γ in B with $\gamma(0) = p(x)$. After Lebesgue (Sect. 4.1), there is an n, so that γ maps each interval of the form $[(k-1)/n, k/n]$ into an open subset of B over which p is trivial. The restriction of γ to $[0, 1/n]$ then has (exactly) one lift starting at x. The endpoint of this lift is used to continue the lift over $[1/n, 2/n]$ and so on. This way, we obtain a lift of γ over the whole $[0, 1]$. In the general case, it is now clear how F should look as a map. If z is a point from Z, then the restriction of F to $I \times \{z\}$ must be the lift of the restriction of H to $I \times \{z\}$ starting at $f(z)$. All that remains is to show the continuity of this map.

We make a preliminary remark. Let J be an open interval in I, let $U \subseteq Z$ be open, and let $H(J \times U)$ be contained in one of the open subsets of B over which p is trivial. (Every point (t, z) from $I \times Z$ has such a neighbourhood.) Up to homeomorphism, the map F on $J \times U$ then has the form (G, H) with a map G from $J \times U$ to the fibre and the map H from $J \times U$ to B as above. The continuity of F on $J \times U$ is equivalent to that of G. By construction, the map G is constant on sets of the form $J \times \{u\}$. The continuity of G on the whole $J \times U$ therefore follows if there is only one t from J such that G is continuous on $\{t\} \times U$. In summary, if the restriction of F to a set of the form $\{t\} \times U$ is continuous, then F is continuous on the whole $J \times U$.

Now we can show the continuity of F. For any point z from Z, let $T(z)$ be the set of t in I such that there is a neighbourhood of (t, z) in $I \times Z$, on which F is continuous. Then it suffices to show the equality $T(z) = I$ for all z. Because I is connected, it is only necessary to know that $T(z)$ is open, closed and not empty. This is what we will now show.

Not empty: we show that 0 is in $T(z)$. For $(0, z)$, there is certainly a neighbourhood of the form $J \times U$ as in the preliminary remark. On $\{0\} \times U$, the map F is given by the continuous initial condition. According to the preliminary remark, the map F is then continuous on $J \times U$.

Open: this follows directly from the definition. If t is in $T(z)$, there is a neighbourhood of the form $K \times V$ of (t, z) on which F is continuous. But then the entire neighbourhood K of t is contained in $T(z)$.

Closed: Let t be in the closure of $T(z)$. For (t, z), there is a neighbourhood $J \times U$ as in the preliminary remark. The set J also contains a point t' from $T(z)$. Therefore, there is a neighbourhood $K \times V$ of (t', z) on which the map F is continuous. Then the map F is continuous on $\{t'\} \times V$, so in particular on $\{t'\} \times (U \cap V)$. According to the preliminary remark, the map F is continuous on the whole $J \times (U \cap V)$. Thus, the t is in $T(z)$. □

We obtain the following result as a first application of the lifting theorem.

Corollary 8.2.3
Let $p \colon X \to B$ be a covering. Then the functor

$$p_* \colon \Pi(X) \longrightarrow \Pi(B)$$

is injective on sets of morphisms, i.e., faithful.

Proof. Let γ and γ' be paths in X from x to x', so that $p\gamma$ and $p\gamma'$ are homotopic with fixed endpoints. Then according to the lifting theorem, the homotopy can be lifted. This provides a homotopy from γ to a lift of $p\gamma'$. The endpoints are fixed here because there are no non-constant paths in the fibre. Due to the uniqueness, the lift of $p\gamma'$ must coincide with γ'. □

Supplement

Another Lifting Theorem The lifting theorem allows the following variant. Let Y be path-connected and locally path-connected. Let $p \colon X \to B$ be a covering. Let $f \colon Y \to B$ be a continuous map and let $y \in Y$ and $x \in X$ be two points with the property $f(y) = p(x)$. Then f has a lift $F \colon Y \to X$ with $F(y) = x$ if and

8.2 The Lifting Theorem

only if the associated algebraic problem

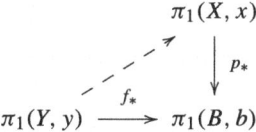

is solvable, i.e., if

$$f_*(\pi_1(Y, y)) \subseteq p_*(\pi_1(X, x))$$

applies. To construct F at a point y', we again choose a path from the base point y to y' and lift its image under f in x. The endpoint of this path then gives the lift $F(y')$. The well-definedness of this map F follows from the algebraic condition: any two paths from y to y' differ only by a homotopy relative endpoints through a loop in y. Its image under f can be lifted to a closed path and, therefore, does not change the endpoint. The continuity of the map F follows from the local path-connectivity of Y. If $y' \in Y$ and U is a sheet of p that contains $F(y')$, then there is a path-connected neighbourhood V of y' with $f(V) \subseteq p(U)$. It suffices to show that $F(V) \subseteq U$ holds. For this, choose a path γ from y to y'. To get a path from y to any point $y' \in V$, we can continue γ with a path γ' that runs entirely in V. A lift of $f(\gamma'\gamma)$ in x results from a lift of $f(\gamma)$ in x followed by a lift of $f(\gamma')$ in $F(y')$. Their endpoint is obviously in U.

Exercises

Exercise 135 Cross and Cross
Find all lifts of the diagonal path

$$\gamma: I \longrightarrow K, \ t \mapsto [(t, t)]$$

in the covering of the Klein bottle K by the torus in Sect. 8.1. Sketch these lifted paths in the torus after the identification on the edge of the square.

Exercise 136 Looking at the End
Let $p: X \to B$ be a covering, where X is path-connected and B is simply-connected. Show that p must be a homeomorphism. (Hint: connect two points in the fibre by a path and lift a null-homotopy of the associated loop in B.)

Exercise 137 Whiplash
Let $p: X \to B$ be a covering. If two points x and x' over b are in the same path component of X, then the subgroups $p_*\pi_1(X, x)$ and $p_*\pi_1(X, x')$ in $\pi_1(B, b)$ are conjugated.

8.3 Fibre Transport

If $p: X \to B$ is a covering, then each fibre $p^{-1}(b)$ is a discrete space, i.e., a set. Using the lifting theorem, we can extend the assignment $b \mapsto p^{-1}(b)$ to a functor

$$M_p: \Pi(B) \longrightarrow \mathbf{Sets}.$$

The objects of $\Pi(B)$ are the points of B, and we assign to a point b the fibre

$$M_p(b) = p^{-1}\{b\}.$$

To the class $[\gamma]$ of a path $\gamma: b \to b'$, we must assign a map from $M_p(b)$ to $M_p(b')$ that will be denoted by $M_p[\gamma]$ or simply $[\gamma]$. This is done by assigning to an element x the endpoint of the (uniquely determined) lift $\tilde{\gamma}$ of γ with start x:

$$M_p[\gamma](x) = \tilde{\gamma}(1).$$

Figure 8.6 indicates why this endpoint only depends on the class of γ. If we lift a homotopy relative to the endpoints so that all paths of the homotopy start in x, then the paths must also have the same endpoint; otherwise, we had a non-constant path in the discrete fibre over b'. This provides the data of a functor. The functor axioms are verified by clever choice of suitable representatives for the identity and the composition. The identities are represented by constant paths, where the lifts can then also be chosen constant. They then give as fibre transport also the identities. When $\gamma: b \to b'$ and $\gamma': b' \to b'$ are two composable paths and $\tilde{\gamma}$ and $\tilde{\gamma}'$ lifts of them with starting points x and x', then $\tilde{\gamma}'\tilde{\gamma}$ is a lift of $\gamma'\gamma$ with start x. Therefore $M_p[\gamma']M_p[\gamma]$ and $M_p[\gamma'\gamma]$ both map the point x to the endpoint of $\tilde{\gamma}'$.

Fig. 8.6 Lifts of paths

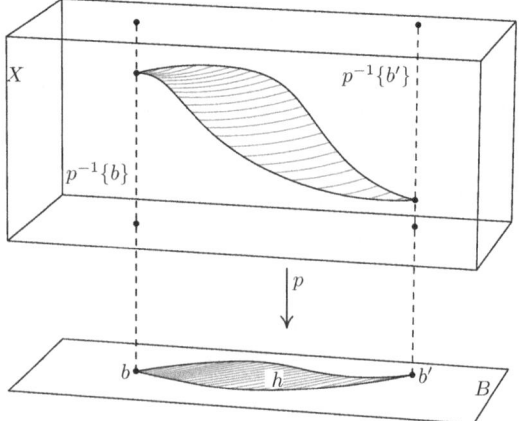

8.3 Fibre Transport

Fig. 8.7 Two three-sheeted coverings of the Klein bottle through itself

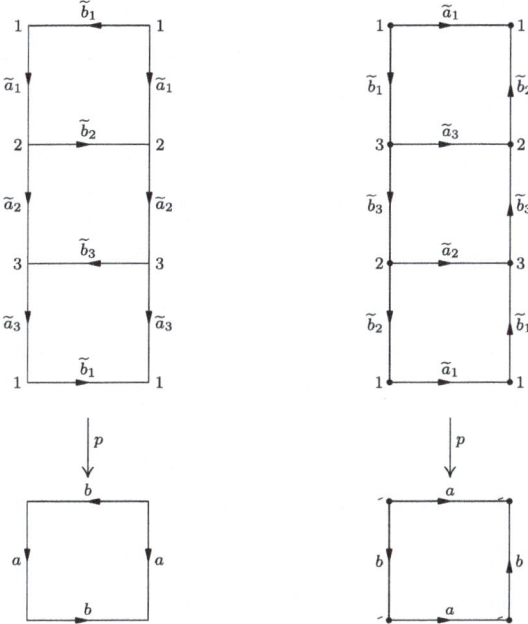

Definition 8.3.1

The functor M_p is the *fibre transport functor* (or *monodromy functor*) of p.

Examples 8.3.2

If $p = \mathrm{pr}_B \colon F \times B \to B$ is a trivial covering, then M_p is isomorphic to the constant functor with value F. Figure 8.7 shows two three-sheeted coverings of the Klein bottle by itself. (The covering to the left was presented differently on the cover of the first German edition of this book.)
The points in the fibre above the initial value of the two loops a, b are numbered. The loops, therefore, act on the fibre $\{1, 2, 3\}$. For the left covering, the fibre transport at this point is described by

$$a \longmapsto (123), \quad b \longmapsto \mathrm{id}$$

in the usual cycle notation: (123) means $1 \mapsto 2$, $2 \mapsto 3$ and $3 \mapsto 1$. In the right covering, we have

$$a \longmapsto (23), \quad b \longmapsto (132)$$

instead.

Definition 8.3.3

A functor from $\Pi(B)$ to the category of sets is also called a $\Pi(B)$-*set*. The $\Pi(B)$-sets form a category. Morphisms between two $\Pi(B)$-sets M and N are the natural transformations. A morphism is also called a $\Pi(B)$-*map*.

In detail, a $\Pi(B)$-map $\Phi \colon M \to N$ does the following. It assigns to each point b from B a map $\Phi(b) \colon M(b) \to N(b)$ in such a way that for every morphism $[\gamma] \colon b \to b'$ the diagram

$$\begin{array}{ccc} M(b) & \xrightarrow{\Phi(b)} & N(b) \\ {\scriptstyle M[\gamma]}\downarrow & & \downarrow{\scriptstyle N[\gamma]} \\ M(b') & \xrightarrow{\Phi(b')} & N(b') \end{array}$$

commutes.

Theorem 8.3.4
The assignment $p \mapsto M_p$ that assigns to each covering of B its fibre transport functor, is itself a functor

$$M \colon \mathbf{Cov}(B) \longrightarrow \Pi(B)\text{-}\mathbf{Sets}.$$

Proof. If f is a morphism from the covering $p \colon X \to B$ to the covering $q \colon Y \to B$, then f maps the fibres $M_p(b)$ and $M_q(b)$ into each other. If $\tilde{\gamma}$ is a lift of γ in $x \in X$ with respect to p, then $f\tilde{\gamma}$ is a lift of γ in $f(x)$ with respect to q. Thus, we have

$$f M_p[\gamma](x) = f\tilde{\gamma}(1) = M_q[\gamma](f(x)),$$

and we obtain a morphism $M_p \to M_q$ of $\Pi(B)$-sets. The functoriality of M also follows. □

It is worth noting here that, by restriction, for each point b of B, we obtain a functor from $\pi_1(B, b)$ into the category of sets that maps the object b to the fibre over b. In other words, the group $\pi_1(B, b)$ acts on the fibre over b through fibre transport. The following results provide more detailed information about this action.

Theorem 8.3.5
Let $p \colon X \to B$ be a covering. For each point x of X, the induced group homomorphism

$$p_* \colon \pi_1(X, x) \longrightarrow \pi_1(B, p(x))$$

(continued)

8.3 Fibre Transport

Theorem 8.3.5 (continued)
injective. The image is the stabiliser of x for the action of $\pi_1(B, p(x))$ on the fibre through x. Two points of this fibre are in the same orbit of this action if and only if they are in the same path-connected component of X.

Proof. The injectivity of p_* on fundamental groups follows immediately from the injectivity of p_* on the fundamental groupoids. If γ is a loop in B at $p(x)$, then $[\gamma]x = x$ is equivalent to the lift of γ starting at x also ending in x. If now $[\gamma]$ is in the stabiliser of x, then the lift of γ is a loop, and $[\gamma]$ is then the image of this loop under p_*. Conversely, if $[\gamma]$ is in the image, then γ is the image of a loop at x. But this is automatically the lift of γ starting at x. Thus, the loop $[\gamma]$ stabilises the point x. If x' is in the orbit of x, say $x' = [\gamma]x$, then x' is the endpoint of the lift of γ starting at x. This lift is then a path from x to x'. Conversely, if x' is a point in the same fibre as x, and there is a path from x to x' in X, then this path is mapped onto a loop γ in B at $p(x)$, whose lift is this path. It follows $[\gamma]x = x'$. □

A particular case of the last theorem is particularly noteworthy.

Corollary 8.3.6
Let $p: X \to B$ be a covering. If X is path-connected, then every fundamental group of B acts transitively on the corresponding fibre.

The results obtained in this section have exciting applications to coverings that arise from group actions. This will be pursued in Sect. 9.3.

Exercises

Exercise 138 Tyres and Transport
Describe the action of the fundamental group of the base on the fibre in the case of the covering p of the Klein bottle by the torus in Sect. 8.1. Also, determine the map induced by p on the fundamental groups.

Exercise 139 Paginated
Show that the number of sheets of a path-connected covering $p: X \to B$ coincides with the index of the subgroup $p_*\pi_1(X, x)$ in $\pi_1(B, p(x))$.

Exercise 140 Pure Braids Again
The n-th pure braid group was defined as the fundamental group of the space $\text{Emb}(n, \mathbb{R}^2)$ of embeddings of a n-element set into the plane. On this space, the symmetric group on n elements acts freely. The quotient is the space $\text{Sub}(n, \mathbb{R}^2)$ of the n-element subsets of the plane. Its

fundamental group is called the *n*-th *(full) braid group*. Show that the *n*-th *(full) braid group* maps surjectively onto the symmetric group, and the kernel is isomorphic to the pure braid group.

Exercise 141 Journey to Jordan
Let $p\colon X \to B$ be a finite covering with $n \geqslant 2$ sheets and non-empty, path-connected spaces X and B. Then there is a loop in B that cannot be lifted to X. (The corresponding group-theoretical result was proven by Jordan, see [Ser03].)

8.4 The Classification Theorem

Let B be a topological space. In this section, we will attempt to describe the category of coverings of B. For this, we can use the functor

$$M\colon \mathbf{Cov}(B) \longrightarrow \Pi(B)\text{-}\mathbf{Sets}$$

that assigns to each covering its fibre transport functor.

▶ **Remark 8.4.1** The functor M is faithful.

If $p\colon X \to B$ is a covering of B, then the set underlying X is the disjoint union of the fibres $M_p(b)$. The natural transformation associated with a map

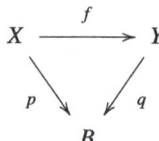

of coverings is just defined by applying f. The question is whether this functor is also full, whether every natural transformation $M_p \to M_q$ can be realised by a morphism f of coverings. If a natural transformation $M_p \to M_q$ is given, there is certainly a map f between the sets underlying X and Y so that the above diagram commutes. A point x is then mapped to the image of x under the map $M_p(p(x)) \to M_q(p(x))$. The only question is whether the map f thus defined is also continuous because then M_f is just this natural transformation. We can answer this using an additional assumption. In Sect. 6.2, we called a topological space Sect. 6.2, we called a topological space *locally path-connected* if every neighbourhood of each of its points contains a path-connected neighbourhood. If $X \to B$ is a covering, then X and B are locally homeomorphic; therefore X is locally path-connected if B is. If the covering is surjective, the converse also holds.

8.4 The Classification Theorem

Theorem 8.4.2
Let B be a locally path-connected topological space. A map f between coverings $p\colon X \to B$ and $q\colon Y \to B$ of B is continuous if and only if the maps $M_p(b) \to M_q(b)$ given by $x \mapsto f(x)$ define a $\Pi(B)$-map.

Proof. One direction is clear: if f is continuous, it induces a $\Pi(B)$-map. Conversely, let f be a map that is compatible with the fibre transport. We must show that f is continuous. So let x be a point of X and $f(x)$ its image. Let V be an open neighbourhood of $f(x)$. By shrinking it, we can achieve that q is a homeomorphism between V and $q(V)$. Let $t\colon q(V) \cong V$ be its inverse. Due to the continuity of p, there is an open neighbourhood U of x with $p(U) \subseteq q(V)$. Because X is locally path-connected, we can achieve that U is path-connected. It is now sufficient to show that $f(U) \subseteq V$ holds. To this end, let x' be another point from U. Then there is a path γ from x to x' in U, and $p\gamma$ is a path from $p(x)$ to $p(x')$ in $p(U) \subseteq q(V)$. According to the definition of fibre transport, we have $M_p[p\gamma](x) = x'$. The compatibility of f with the fibre transport implies then

$$M_q[p\gamma](f(x)) = f(M_p[p\gamma](x)) = f(x').$$

It remains to show that the fibre transport $M_q[p\gamma](f(x))$ lands in V. But $tp\gamma$ is a lift of $p\gamma$ along q with start $f(x)$, and the endpoint of it lies in V. $\square$

As explained before, this leads to the answer to our question about continuity.

Corollary 8.4.3
Let B be a locally path-connected topological space. Then the functor

$$M\colon \mathbf{Cov}(B) \longrightarrow \Pi(B)\text{-}\mathbf{Sets}$$

that assigns to each covering its fibre transport functor, is fully faithful. In particular, two coverings of B are isomorphic if and only if they have isomorphic fibre transports.

Next, we describe the isomorphism classes of coverings of B using the fibre transport. To an isomorphism class of coverings of B, there belongs an isomorphism class of $\Pi(B)$-sets. The question now arises whether all $\Pi(B)$-sets can be realised as transport functors up to isomorphism, i.e., whether the above functor is essentially surjective. This can be answered using another assumption.

Definition 8.4.4 (Simply-Connected! Semi-locally)

A topological space B is called *semi-locally simply-connected*, if B is covered by path-connected, open sets U such that the homomorphism

$$\pi_1(U, u) \longrightarrow \pi_1(B, u)$$

induced by the inclusion is constant for all u from U. An equivalent formulation is: any two paths from u to u' in U define the same morphism in $\Pi(B)$.

Example 8.4.5
The condition is met when every point has a simply-connected neighbourhood. This is true for surfaces or, more generally, for manifolds.

Theorem 8.4.6 (Main Theorem of Covering Theory)
Let B be a locally path-connected and semi-locally simply-connected topological space. Then the functor

$$M \colon \mathbf{Cov}(B) \longrightarrow \Pi(B)\text{-}\mathbf{Sets}$$

that assigns to each covering its fibre transport functor is an equivalence of categories.

Proof. It has already been shown that the functor M is fully faithful (see the corollary before). Due to Theorem 7.1.14, it remains only to be shown that M is also essentially surjective. Let $S \colon \Pi(B) \to \mathbf{Sets}$ be a $\Pi(B)$-set. Then a covering

$$p_S \colon X_S \longrightarrow B$$

must be constructed whose fibre transport functor is isomorphic to S. For this purpose, let

$$X_S = \coprod_{b \in B} S(b)$$

and p_S be constant on $S(b)$ with value b. Let b be a point of B and U a path-connected open neighbourhood of b in B such that $\pi_1(U, b) \to \pi_1(B, b)$ is trivial. According to the assumption, such neighbourhoods cover all of B. If then γ is a path in U that connects b with a point u, then

$$S[\gamma] \colon S(b) \longrightarrow S(u)$$

8.4 The Classification Theorem

does not depend on $[\gamma]$ because any two paths define the same morphism in $\Pi(B)$ and S is determined by this. This allows us to define a map

$$j(b, U) \colon U \times S(b) \longrightarrow p_S^{-1}(U)$$

through $(u, x) \mapsto S[\gamma](x)$. This map is bijective. Let

$$h(b, U) \colon p_S^{-1}(U) \longrightarrow U \times S(b)$$

be the inverse map. A subset of X_S is called *open* if its image after intersection with $p_S^{-1}(U)$ under $h(b, U)$ is open for every (b, U). This defines a topological structure on X_S for which all $h(b, U)$'s are homeomorphisms. The map p_S is then continuous, and the $h(b, U)$'s provide trivialisations, showing that p_S is a covering. It remains to be shown that the fibre transport of $p_S \colon X_S \to B$ is isomorphic to S. First, we note that the fibre transport and S agree on objects. The fibre of p_S over b is the summand $S(b)$ of X_S and is identified with $S(b)$ by the canonical injection. If now γ is a path in B and $\tilde{\gamma}$ is the lift of it along p_S starting at x, then $\tilde{\gamma}(1)$ should be given by $S[\gamma](x)$. It is sufficient to verify this for paths γ that run in a set U as above and have b as a starting point because every path is a composition of such paths. But over U, there is now a trivialisation $j(b, U)$, and for the lift we thus have

$$\tilde{\gamma}(t) = j(b, U)(\gamma(t), x).$$

The end of this lift is $S[\gamma](x)$ by definition of $j(b, U)$. □

It would also be easy to provide an inverse equivalence. To do this, we would only have to consider that the assignment $S \mapsto (p_S \colon X_S \to B)$ from the proof of the theorem is functorial in S. We will not be elaborate on this here. The following section will serve to explain this theorem by interpreting it and varying it—under further assumptions about the connectivity of the spaces.

Exercises

Exercise 142 Small Stuff
Show that the two self-coverings of the Klein bottle from Sect. 8.3 are not isomorphic to each other by examining the corresponding fibre transports.

Exercise 143 This Happens
How many isomorphism classes of connected coverings of $\mathbb{R}P^2 \times \mathbb{R}P^2$ are there? How many morphisms are there between them?

Exercise 144 Base Change
Let $f, g \colon B' \to B$ be two homotopic maps between locally path-connected and semi-locally simply-connected topological spaces. Show for each covering $p \colon X \to B$, that the base changes of p along f and g are isomorphic to each other.

8.5 Topological Galois Theory

It turns out that the covering theory of topological spaces has many formal similarities with the algebraic Galois theory of field extensions. This becomes particularly clear when the spaces are sufficiently connected. In this section, therefore, the base B is always a connected, locally path-connected and semi-locally simply-connected topological space. If B is a path-connected space, then not only are $\Pi(B)$ and $\pi_1(B, b)$ equivalent, but also the categories of $\Pi(B)$-sets and $\pi_1(B, b)$-sets for each point b. This follows from general reasons but can also be explicitly justified for this situation. It is necessary to construct an inverse equivalence to the restriction functor from the category of $\Pi(B)$-sets to the category of $\pi_1(B, b)$-sets. Let N be a $\pi_1(B, b)$-set. By assumption, for each point c there is a path φ_c from c to b. The functor belonging to N should then be given by $N(c) = N$ on objects. The $[\gamma : c \to d]$ associated map $N(c) \to N(d)$ is given by the action of $[\varphi_d \gamma \varphi_c^{-1}]$. We can verify that this defines a $\Pi(B)$-set that also functorially depends on N. If $N = M(b)$ for a $\Pi(B)$-set M, then the $\Pi(B)$-set belonging to N is isomorphic to M. A $\Pi(B)$-bijection is given by the paths φ_c.

Corollary 8.5.1
Let B be a connected, locally path-connected and semi-locally simply-connected topological space. If b is an arbitrary point in B, then the fibre transport functor provides an equivalence

$$\mathbf{Cov}(B) \longrightarrow \pi_1(B, b)\text{-}\mathbf{Sets}$$

of categories.

In this equivalence, the connected coverings are mapped to the transitive $\pi_1(B, b)$-sets. Each such is equivalent to a homogeneous $\pi_1(B, b)$-set of the form $\pi_1(B, b)/H$ for a subgroup H of $\pi_1(B, b)$.

Definition 8.5.2
The *orbit category* $\mathbf{Orb}(G)$ of a group G has as objects the homogeneous G-sets G/H and as morphisms the G-maps.

Corollary 8.5.3
Let B be a connected, locally path-connected and semi-locally simply-connected topological space. If b is any point in B, there is an equivalence between the category of connected coverings of B and the orbit category of $\pi_1(B, b)$.

8.5 Topological Galois Theory

In the following, only coverings should be considered whose total spaces are not empty.

Examples 8.5.4
Let B be a locally path-connected and simply-connected topological space. Then every fundamental group is trivial and consequently also their orbit category; there is a unique object and a unique morphism, namely the identity of the object. This means that every connected covering of B isomorphic to id_B. In other words, every connected covering of B is a homeomorphism.

Let B be a connected, locally path-connected and semi-locally simply-connected topological space, whose fundamental groups are isomorphic to the group $G = \mathbb{Z}/2$. Examples are $\mathbb{RP}^n$ for $n \geqslant 2$. Then the orbit category has two objects $G/1$ and $G/G = \star$. There are two morphisms that are not identities: the automorphism of $G/1$ that swaps the two elements, and the map $G/1 \to G/G$. The space B, therefore, has up to isomorphism a unique non-trivial connected covering; its deck transformation group has order 2, like $S^n \to \mathbb{RP}^n$ for $n \geqslant 2$.

Let B be a connected, locally path-connected and semi-locally simply-connected topological space whose fundamental groups are isomorphic to $\mathbb{Z}$. An example is the circle S^1. Then the objects of the orbit category are given by $\mathbb{Z}/n$ for all $n = 0, 1, 2, 3, \ldots$. We have $\mathbb{Z}/0 \cong \mathbb{Z}$ and $\mathbb{Z}/n$ has exactly n elements when $n \geqslant 1$. The corresponding coverings of $B = S^1$ are $e_n \colon S^1 \to S^1$ with $e_n(z) = z^n$ for $n \geqslant 1$ and the exponential map $e_0 \colon \mathbb{R} \to S^1$ for $n = 0$. Every connected covering is isomorphic to one of these. We can also easily consider how many morphisms there are between these coverings. A morphism $e_m \to e_n$ exists if and only if m is a multiple of n. (And exactly n, when $n \geqslant 0$.) The automorphism group of e_n is isomorphic to $\mathbb{Z}/n$, since $\mathbb{Z}$ is abelian. We also recommend taking a look at a non-abelian example. The symmetric group acting on three elements is suitable. There are six objects in the orbit category, of which three are isomorphic to each other. (This redundancy of the orbit category is preferable to an arbitrary choice of isomorphism classes.) These three objects, although not trivial, have, by the way, trivial automorphism groups. This is generally a good opportunity to return our attention to the automorphism groups.

We recall the Weyl groups from the supplement in Sect. 5.2. If H is a subgroup of a group G, then H is normal in its normaliser NH, and the quotient NH/H is the Weyl group WH of H in G. The significance of these groups is explained by the fact that the automorphism group of the G-set G/H is isomorphic to the Weyl group of H in G. The class nH corresponds to the automorphism $gH \mapsto gn^{-1}H$.

Corollary 8.5.5
Let B be a connected and locally path-connected topological space. Let $p \colon X \to B$ be a connected covering of B. The automorphism group $\mathrm{Aut}(p)$ is isomorphic to the Weyl group of $p_*\pi_1(X, x)$ in $\pi_1(B, p(x))$, for any point x from X.

With this result, the automorphism groups of connected coverings are described by the fundamental groups. How do they act on the total space of the covering? For this, we first need a technical result that does not use the classification theorem. Accordingly, some hypotheses can be omitted.

Theorem 8.5.6
Let $p\colon X \to B$ be a covering. If X is connected, then the deck transformation group $\mathrm{Aut}(p)$ acts freely on X, and for every subgroup H of $\mathrm{Aut}(p)$, the map $f\colon X \to X/H$ is a covering. If X is also locally path-connected, then the induced map $q\colon X/H \to B$ is also a covering, and f is a morphism from p to q.

Proof. Let h be an automorphism of p. Then h and id_X are lifts of p along p. So, if there is a point x in X with $h(x) = x$, then $h = \mathrm{id}_X$ due to the uniqueness of such lifts. If U is a neighbourhood of x that is homomorphically mapped by p, and x' is in $U \cap hU$ for an automorphism h of p, then x' and $h(x')$ are in hU and have the same image in B. So they agree, which implies $h = \mathrm{id}_X$. If X is locally path-connected, then so is B. Therefore, the space B is covered by path-connected open subsets U over which p is trivial. It suffices to show that q is also trivial over these subsets. Because U is path-connected, there is a trivialisation $p^{-1}(U) \cong U \times \pi_0(p^{-1}U)$, and the automorphisms act on it by permuting the components, thus only on the second factor. But this already implies that we have a homeomorphism $q^{-1}(U) \cong U \times (\pi_0(p^{-1}U)/H)$. □

If B is now a connected and locally path-connected topological space, then we can factorise every connected covering $p\colon X \to B$ as a composition

$$X \longrightarrow X/H \longrightarrow B,$$

of two coverings, where H is any subgroup of $\mathrm{Aut}(p)$. In particular, for $H = \mathrm{Aut}(p)$, the question arises again whether the continuous map $q\colon X/\mathrm{Aut}(p) \to B$ is a homeomorphism. This is not always true. The following result characterises the positive cases.

Theorem 8.5.7
Let $p\colon X \to B$ be a connected covering of the connected and locally path-connected space B. Then the following are equivalent:

(a) The map $q\colon X/\mathrm{Aut}(p) \to B$ is a homeomorphism.
(b) The group $\mathrm{Aut}(p)$ acts transitively on each fibre.
(c) The group $\mathrm{Aut}(p)$ acts transitively on a fibre.
(d) The group $p_*\pi_1(X,x)$ is normal in $\pi_1(B, p(x))$ for some x.
(e) The group $p_*\pi_1(X,x)$ is normal in $\pi_1(B, p(x))$ for all x.

8.5 Topological Galois Theory

Proof. As for (a) ⇒ (b): for each fibre F, the set $F/\mathrm{Aut}(p)$ of orbits consists of only one point. Thus, the action is transitive. As for (b) ⇒ (a): if $\mathrm{Aut}(p)$ acts freely and transitively on X, then $X/\mathrm{Aut}(p) \to B$ is a continuous open bijection, thus a homeomorphism. For each point b, according to the classification theorem, the group $\mathrm{Aut}(p)$ is isomorphic to the automorphism group of the fibre as a $\pi_1(B, b)$-set. By Theorem 8.3.5, the fibre is, as a $\pi_1(B, b)$-set, isomorphic to $\pi_1(B, b)/p_*\pi_1(X, x)$ for each point x above b. The equivalence of (b) and (e) now follows from the fact that the automorphism group of it is the Weyl group of $p_*\pi_1(X, x)$ in $\pi_1(B, b)$. It acts transitively if and only if $p_*\pi_1(X, x)$ is normal in $\pi_1(B, b)$. □

Definition 8.5.8

A covering with the properties mentioned in Theorem 8.5.7 is called *regular* or *normal* or a *Galois covering*. These correspond to the Galois extensions in algebra.

Examples 8.5.9

We once again consider the covering of the Klein bottle by the torus from Sect. 8.1. There is an obvious deck transformation f that half-turns the left square and then mirrors it along the vertical axis to the right square. This transformation maps each element of a fibre to the other element of the fibre. So, this covering is regular. In the case of the self-covering of the Klein bottle in the left part of the diagram in Sect. 8.3, the map f just described can be used to move the squares one up, with the upper square being brought down unmirrored. On the fibre, this corresponds to a deck transformation, thus to the cycle (132), and the action is transitive. Because the lifts of b are closed, we have $[b] = 1$ in the quotient $\pi_1(K)/p_*(\pi_1(X))$. This group is thus cyclically generated by $[a]$. The relation $[a]^3 = 1$ can be recognised at the loop $\tilde{a}_3\tilde{a}_2\tilde{a}_1$.

There are universal examples of Galois coverings that are comparable with the algebraic closures in algebra. To describe these, let B be again a connected, locally path-connected and semi-locally simply-connected topological space. Then according to the classification theorem, there are coverings $p\colon X \to B$ on whose fibres the fundamental groups act freely and transitively. After choosing a point b, for example, the functor $? \mapsto \mathrm{Mor}_{\Pi(B)}(b, ?)$ is a $\Pi(B)$-set that belongs to such a covering. With Theorem 8.3.5, we get the following result:

▶ **Remark 8.5.10** The fundamental groups of the base act freely and transitively on the fibres if and only if X is simply-connected if and only if it does not have non-trivial coverings itself.

Definition 8.5.11

Any covering of B with a simply-connected total space X is called a *universal covering* of B.

Fig. 8.8 The universal covering of $S^1 \vee S^1$

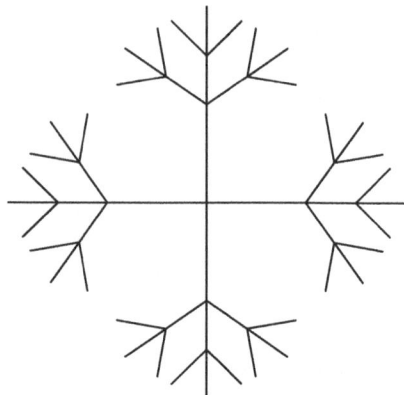

Fig. 8.9 The universal covering of the Klein bottle

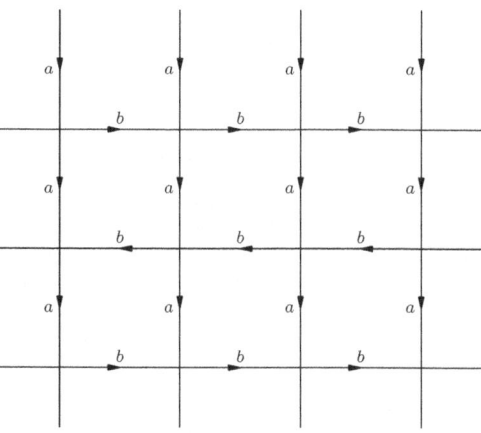

Any two such coverings are isomorphic to each other; however, in general, there is no distinguished isomorphism since the deck transformation groups are isomorphic to the fundamental groups of B. Universal coverings are Galois coverings.

Examples 8.5.12
The exponential map $\mathbb{R} \to S^1$ and the identification $S^n \to \mathbb{R}P^n$ for $n \geq 2$ are universal coverings. The universal covering of the one-point union $S^1 \vee S^1$ of two circles is an infinite graph, a section of which is depicted in Fig. 8.8. The vertices of the graph are mapped to the intersection of the circles. From each of them, four edges originate. They belong to the different directions originating from the intersection of the two circles in $S^1 \vee S^1$.
The universal covering space of the Klein bottle is the plane $\mathbb{R}^2$, which is divided into squares as shown in Fig. 8.9.

Now, we should explain the universal property of universal coverings. We use the notation $\text{Hom}_B(X, Y)$ for the set of morphisms from a covering $X \to B$ to a covering $Y \to B$.

Theorem 8.5.13 (The Universal Pointed Covering)
Let B be a connected, locally path-connected and semi-locally simply-connected topological space. Let $p\colon X_u \to B$ be a universal covering of B and $q\colon Y \to B$ any covering. Then every point x from X_u over b from B provides a bijection

$$\mathrm{Hom}_B(X_u, Y) \longrightarrow q^{-1}(b), \quad f \longmapsto f(x).$$

This bijection is natural in coverings $q\colon Y \to B$ over B.

Proof. If $G = \pi_1(B, b)$, then the classification theorem provides a bijection

$$\mathrm{Hom}_B(X_u, Y) \longrightarrow \mathrm{Abb}_G(p^{-1}(b), q^{-1}(b)), \quad f \longmapsto f\mid p^{-1}(b).$$

In the universal covering, the G-set $p^{-1}(b)$ is isomorphic to $G/1$. Furthermore, there is a unique G-bijection that maps x to 1. This provides a natural bijection

$$\mathrm{Abb}_G(p^{-1}(b), q^{-1}(b)) \longrightarrow \mathrm{Abb}_G(G/1, q^{-1}(b)) \cong q^{-1}(b).$$

The composition of these natural bijections is the one mentioned in the theorem. □

The inclined reader will try to formulate the universal property as a statement about the pullback $X_u \times_B Y$. However, the result can also be interpreted in other ways. Let B be a connected, locally path-connected and semi-locally simply-connected topological space. The fibre transport functor, which assigns to each covering of B its fibre over b with its π_1-action, is an equivalence of categories. If now $p\colon X_u \to B$ is a universal covering, then each point x provides a natural isomorphism between the fibre functor and the functor $\mathrm{Hom}_B(X_u, ?)$. The latter has the advantage of not depending on the choice of b. For each covering $q\colon Y \to B$, the set $\mathrm{Hom}_B(X_u, Y)$ is naturally an $\mathrm{Aut}(p)$-set through $g(f) = fg^{-1}$.

Corollary 8.5.14 (Main Theorem of Topological Galois Theory)
Let B be a connected, locally path-connected and semi-locally simply-connected topological space. Let $p\colon X_u \to B$ be a universal covering of B. Then the functor

$$\mathrm{Hom}_B(X_u, ?)\colon \mathbf{Cov}(B) \longrightarrow \mathrm{Aut}(p)\text{-}\mathbf{Sets}$$

that assigns to each covering $q\colon Y \to B$ the set $\mathrm{Hom}_B(X_u, Y)$ is an equivalence of categories.

Proof. Let $G = \pi_1(B, b)$ for a point b from B. According to the main theorem of covering theory, we have

$$\mathrm{Aut}(p) \cong \mathrm{Abb}_G(p^{-1}(b), p^{-1}(b)).$$

A point x from X_u over b provides an isomorphism

$$\mathrm{Abb}_G(p^{-1}(b), p^{-1}(b)) \cong \mathrm{Abb}_G(G/1, G/1) \cong W1 \cong G.$$

In total, this results in an isomorphism $\mathrm{Aut}(p) \cong \pi_1(B, b)$. This isomorphism depends on the choice of x. It induces an isomorphism

$$\mathrm{Aut}(p)\text{-}\mathbf{Sets} \longrightarrow \pi_1(B, b)\text{-}\mathbf{Sets}.$$

This isomorphism can now be used to compare the functor from the statement of the theorem with the fibre functor to b. This is done using the natural bijection

$$\mathrm{Hom}_B(X_u, Y) \cong q^{-1}(b), \quad f \longmapsto f(x),$$

that also depends on x. The statement now follows from the main theorem of covering theory, once we have shown that the natural bijection is compatible with the $\mathrm{Aut}(p)$- and $\pi_1(B, b)$-actions in the way described above. If $f \colon X_u \to Y$ is a morphism and g is an automorphism of p, then on the one hand, the morphism fg^{-1} maps the point x to $f(g^{-1}(x))$. On the other hand, the corresponding element $[\gamma]$ of the fundamental group acts on $q^{-1}(b)$ and maps $f(x)$ to $M_q[\gamma]f(x)$. Due to the compatibility of f with the fibre transport, we have $M_q[\gamma]f(x) = f(M_p[\gamma]x)$. The relation $g^{-1}(x) = M_p[\gamma]x$ will be explained on another occasion, namely in the proof of Theorem 9.3.5. □

An inverse to the equivalence in the preceding corollary is not difficult to describe, by the way. This will be done in Sect. 9.2.

Example 8.5.15
With tools from linear algebra (see [Brö03, Sec. IX.3]), it is easy to construct a double covering

$$\mathrm{SU}(2) \longrightarrow \mathrm{SO}(3)$$

that is a surjective morphism of groups with a kernel of order two. The topological group $\mathrm{SU}(2)$ is homeomorphic to S^3 as a space, so it is simply-connected. This is the universal covering of $\mathrm{SO}(3)$. The fundamental group of $\mathrm{SO}(3)$ is therefore isomorphic to $\mathbb{Z}/2$. Indeed, the space $\mathrm{SO}(3)$ is homeomorphic to $\mathbb{R}P^3$. The group $\mathrm{SO}(3)$ contains a subgroup isomorphic to the alternating group A_5, the group of symmetries of a dodecahedron or icosahedron. Let G be its pre-image in $\mathrm{SU}(2)$. Then the group G acts freely on $\mathrm{SU}(2) \cong S^3$, and S^3 is the universal covering of the quotient $P = S^3/G$. This is the famous *Poincaré sphere* whose fundamental group is isomorphic to G. According to the main theorem, coverings $q \colon Y \to P$ are classified by the G-set of continuous maps $S^3 \to Y$ over P.

8.5 Topological Galois Theory

Example 8.5.16
Let $T = (S^1)^n$ be the n-dimensional torus. Its fundamental group is isomorphic to $\mathbb{Z}^n$, and a universal covering is given by the n-fold exponential function $\mathbb{R}^n \to T$. According to the main theorem, coverings $q: Y \to T$ are classified by the $\mathbb{Z}^n$-set of continuous maps $\mathbb{R}^n \to Y$ over T.

Supplements

Algebraic Galois Theory The equivalence we gave in the main theorem of topological Galois theory is intrinsic in several respects. It does not depend on the choice of points, and—what is even more decisive—it does not refer to the fundamental group or the fundamental groupoid. Because algebra does not know paths, the main theorem of algebraic Galois theory is often given in this form. If $\overline{k}$ is an algebraic closure and $G = \mathrm{Aut}(\overline{k}|k)$ is the Galois group of k, then to every field extension $K|k$ we can assign the G-set of k-embeddings $K \to \overline{k}$. Conversely, every G-set M yields the k-algebra $\mathrm{Abb}_G(M, \overline{k})$; if $M \cong G/H$ is transitive, then

$$\mathrm{Abb}_G(M, \overline{k}) \cong \mathrm{Abb}_G(G/H, \overline{k}) \cong \overline{k}^H$$

is the fixed field. Under suitable conditions, these constructions are equivalences of categories. We refer to [Dre95] for a concise presentation of the main results of algebraic Galois theory from this standpoint.

The Classification Theorem in the Pointed Setting In topology, it is often helpful to use the category of pointed spaces instead of the category of topological spaces so that certain constructions become functorial. A *pointed* topological space is a pair (X, x), consisting of a topological space X and a point x from X. A morphism $f: (X, x) \to (Y, y)$ is a continuous map $f: X \to Y$ with $f(x) = y$. The pointed topological spaces form a category in an obvious way. It is isomorphic to the category of spaces under the one-point space $\star$, where (X, x) is identified with the map $\star \to X$ with value x. On the category of pointed topological spaces, we have the functor π_1. A morphism $p: (X, x) \to (B, b)$ of pointed topological spaces is a *pointed covering* if $p: X \to B$ is a covering. If $p: (X, x) \to (B, b)$ is a pointed covering of (B, b), then the pointed fibre $(p^{-1}(b), x)$ is a pointed $\pi_1(B, b)$-set. (A *pointed G-set* is a pair (M, m), consisting of a G-set M and an element m. The element m does not have to be a G-fixed point.) The classification theorem for pointed and sufficiently connected base spaces (B, b) now states in its pointed version that the fibre transport functor, which assigns to a pointed covering $p: (X, x) \to (B, b)$ the pointed $\pi_1(B, b)$-set $(p^{-1}(b), x)$, is an equivalence of categories. From the observation that there is at most one pointed G-map between the pointed G-sets $(G/H, 1/H)$ and $(G/H', 1/H')$, and this exists if and only if $H \subseteq H'$ holds, it follows that the category of connected pointed coverings of (B, b) is equivalent to the category of subgroups of $\pi_1(B, b)$ and their inclusions. An equivalence assigns to $p: (X, x) \to (B, b)$ the subgroup $p_*\pi_1(X, x)$.

Exercises

Exercise 145 That's It
Investigate whether the self-covering of the Klein bottle on the right side of the figure in Sect. 8.3 is regular. Show that there are no other path-connected three-sheeted coverings of the Klein bottle except these two.

Exercise 146 More than 99 Red Balloons
Describe a universal covering of the one-point union $S^1 \vee S^2$.

Exercise 147 Despite Orientation
How many two-sheeted path-connected coverings of the pretzel surface are there?

Exercise 148 What They Can Do
Choose from any algebra book its main theorem of Galois theory, translate it (as literally as possible) into a result of covering theory, and infer it from the topological Galois equivalence.

Bundles and Fibrations 9

Bundles are an essential class of maps that generalise coverings. The fibre of a general bundle must no longer be discrete but can be any topological space. Important examples of fibre bundles are principal bundles and vector bundles, whose fibres are topological groups and vector spaces, respectively. This chapter introduces these concepts.

9.1 Fibre Bundles

It is often helpful to consider a continuous map $p\colon X \to B$ as a 'continuous' family

$$(p^{-1}(b) \mid b \in B)$$

of spaces, namely their fibres. This perspective is the focus of this chapter. We should again formulate it a more boldly: a continuous map p provides a 'continuous' map $b \mapsto p^{-1}(b)$ from B to a 'space of all spaces'. We will not attempt to make this precise, but it will stimulate our imagination at some points. For example, suppose

$$\begin{array}{ccc} X' & \longrightarrow & X \\ p' \downarrow & & \downarrow p \\ B' & \xrightarrow{f} & B \end{array}$$

is a pullback so that p' is a base change of p. Then the fibre of p' over b' is canonically homeomorphic to the fibre of p over $f(b')$. The map into the 'space of all spaces' corresponding to p' therefore results from that of p by composition with f (and a canonical homeomorphism). This explains the role of the base change in this perspective.

The simplest situation arises when X is a product $B \times F$ and p is the projection. Then all fibres are (canonically) homeomorphic (to F).

Definition 9.1.1

A continuous map $p\colon X \to B$ is called *trivial with typical fibre F* if there is a homeomorphism $h\colon X \to B \times F$ such that the diagram

$$\begin{array}{ccc} X & \xrightarrow{h} & B \times F \\ {\scriptstyle p}\downarrow & \swarrow {\scriptstyle \mathrm{pr}_B} & \\ B & & \end{array}$$

commutes. Such a homeomorphism h is then called a *trivialisation* or a *chart* of p. A map is trivial if and only if it is a base change of the map $F \to \star$.

Example 9.1.2
The tangent bundle of the circle line is the subspace

$$TS^1 = \{(z, v) \mid v \perp z\} \subseteq S^1 \times \mathbb{C}$$

together with the projection $(z, v) \mapsto z$ onto the first coordinate in S^1. This map is trivial, with typical fibre $\mathbb{R}$. A trivialisation is given by

$$TS^1 \longrightarrow S^1 \times \mathbb{R}, \quad (z, v) \mapsto (z, \frac{v}{iz})$$

as shown in Fig. 9.1.

Any two trivialisations $X \cong B \times F$ of a map $X \to B$ differ by an automorphism of $B \times F$ that in the first component is the identity and in the second component is given by a continuous map

$$\Psi \colon B \times F \longrightarrow F.$$

Fig. 9.1 The tangent bundle of the circle S^1

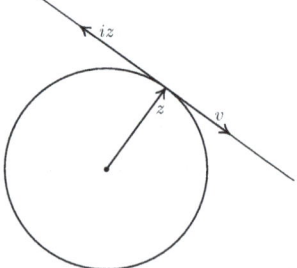

9.1 Fibre Bundles

For each b, the map $f \mapsto \Psi(b, f)$ is an automorphism of F. Therefore, the map adjoint to Ψ is a continuous map

$$B \longrightarrow \mathrm{Aut}(F), \quad b \longmapsto (f \mapsto \Psi(b, f)).$$

Conversely, if F is locally compact, then we can use any such map to modify trivialisations.

Examples 9.1.3
A continuous map does not need to be trivial, even if the fibres are all homeomorphic. A perhaps less interesting example is the identical map from $\mathbb{R}$ to $\mathbb{R}$, where the source has the discrete and the target the usual topology. A somewhat more interesting example is the map $q \colon S^1 \to S^1$ with $q(z) = z^2$. This is not trivial. At least two arguments should be known for this. First, if the map were trivial then the source S^1 would be homeomorphic to two copies of S^1, and that is a contradiction. Second: if $p \colon X \to B$ is trivial, and the typical fibre is not empty, then there is a *section* of p, i.e., a continuous map $s \colon B \to X$ with $ps = \mathrm{id}_B$. If now the map q had a section, then the degree of the identity map would be even. That also gives a contradiction. The map q is indeed not trivial, but it is locally trivial in the following sense.

Definition 9.1.4

A continuous map $p \colon X \to B$ is called *locally trivial with typical fibre F* if there is a cover of B such that the base change from p to each cover set is trivial with typical fibre F. In other words, for each of these cover sets U, there is a diagram

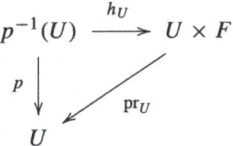

where h is a homeomorphism.

Examples 9.1.5
Every trivial map is locally trivial. The converse is only sometimes true, as we have already seen by the maps derived from coverings. Coverings are locally trivial with discrete fibre. Another example of a locally trivial map is the Möbius strip over the circle line, shown in Fig. 9.2. The typical fibre is an interval. If we remove a point from the base S^1, the restricted map becomes trivial. The map itself is not trivial because any two sections always meet in at least one point according to the intermediate value theorem.

Also, the map from the Klein bottle to the circle (see Fig. 9.3) is only locally trivial with typical fibre S^1 because the fundamental group of the Klein bottle is not isomorphic to that of the torus (see the exercise on the Klein bottle in Sect. 7.3).

Definition 9.1.6

A set of charts (U, h_U) for base changes from p to open subsets $U \subseteq B$ is an *atlas* of p if the sets U cover all of B. Every locally trivial map has an atlas. Once an atlas is chosen, its charts can be compared when restricted to the

Fig. 9.2 The Möbius strip is locally trivial over the circle line

Fig. 9.3 The Klein bottle is locally trivial over the circle line

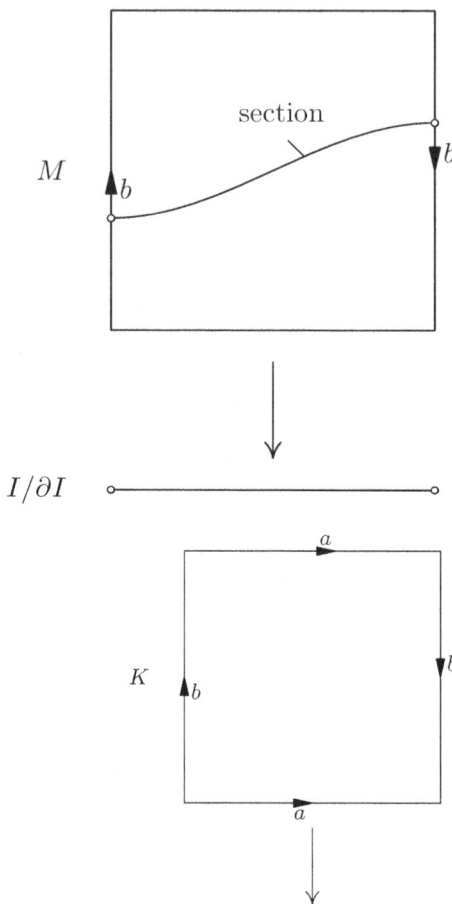

intersections $U \cap V$. This provides continuous maps

$$\Phi_{UV} \colon U \cap V \longrightarrow \operatorname{Aut}(F)$$

that describe the second component of the upper line in the diagram

$$(U \cap V) \times F \xrightarrow{h_V^{-1}} p^{-1}(U \cap V) \xrightarrow{h_U} (U \cap V) \times F$$
$$\searrow \qquad \downarrow \qquad \swarrow$$
$$U \cap V.$$

9.1 Fibre Bundles

The first component is the identity. As a formula, we get

$$\Phi_{UV}(b)(f) = \mathrm{pr}_F h_U h_V^{-1}(b, f).$$

These are the *transition functions* or *chart changes* of the atlas.

The transition functions describe which symmetries of F are used to switch from one chart to another. In some situations, we already know certain symmetries of F and only want to allow these for the transition functions. This leads to the following notion.

Definition 9.1.7

Let G be a topological group that acts on the space F *effectively* (or *faithfully*), i.e., the group homomorphism

$$G \longrightarrow \mathrm{Aut}(F)$$

adjoint to the action should be injective. (It does not have to be an embedding.) Let furthermore $p \colon X \to B$ be a locally trivial map with typical fibre F. An atlas of p is a *G-atlas* if there are continuous maps

$$g_{UV} \colon U \cap V \longrightarrow G$$

so that the transition functions of the atlas are given by the composite of the g_{UV} with the group homomorphism $G \to \mathrm{Aut}(F)$:

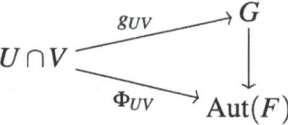

Due to the effectiveness of the action, the g_{UV} are uniquely determined by the transition functions; so the only question is whether the image of all Φ_{UV} is in the image of $G \to \mathrm{Aut}(F)$, and if this is the case, whether the resulting maps g_{UV} are then continuous. Two G-atlases are called *equivalent* if their union is again a G-atlas. Every equivalence class of atlases contains a maximal atlas.

Definition 9.1.8

A *G-fibre bundle with typical fibre F* is a locally trivial map $p \colon X \to B$ with typical fibre F together with a maximal G-atlas of p. The group G is often called the *structure group* of the fibre bundle. The space X is called the *total space*, the space B the *base* and p the *projection*.

Note that it is not enough to know G and F to speak of G-fibre bundles with typical fibre F; we must also know the action of G on F, i.e., the homomorphism $G \to \text{Aut}(F)$. Every G-fibre bundle with typical fibre F is locally trivial with typical fibre F. Conversely, every locally trivial map with typical fibre F can be considered as a fibre bundle with typical fibre F and structure group $\text{Aut}(F)$. This works if $\text{Aut}(F)$ is a topological group that acts continuously on F, for example, if F is a compact Hausdorff space.

Examples 9.1.9
The above example of the Klein bottle over the circle is an example of a $\mathbb{Z}/2$-fibre bundle with typical fibre S^1, where

$$\mathbb{Z}/2 \longrightarrow \text{Aut}(S^1)$$

maps the element -1 to the reflection $z \mapsto \bar{z}$. Accordingly, the Möbius strip is a $\mathbb{Z}/2$-fibre bundle with typical fibre being an interval, on which the group $\mathbb{Z}/2$ acts by reflection at the midpoint.

Supplement

Bundle Constructions Let G be a topological group that acts effectively on F. Furthermore, let there be a cover of a space B given and for each pair of sets U and V in the cover a continuous function

$$g_{UV} : U \cap V \longrightarrow G$$

satisfying the co-cycle conditions (see the subsequent exercise). We can ask whether there is a G-fibre bundle with typical fibre F whose transition functions coincide with the g_{UV} in $\text{Aut}(F)$. This is indeed the case and will be briefly indicated here: as total space X, take the topological sum of the products $U \times F$ for each U from the cover and introduce the following equivalence relation: for all $b \in U \cap V$ and all $f \in F$, set

$$(b, f)_V \sim (b, g_{UV}(b)f)_U.$$

Due to the cocycle condition, this is an equivalence relation. The map from X to B that sends $[(b, f)_U]$ to b has the desired properties.

Exercises

Exercise 149 Cocycle
The transition functions $g_{UV} : U \cap V \to G$ of a G-fibre bundle fulfil $g_{UV} g_{VW} g_{WU} = \text{id}$ on $U \cap V \cap W$. In particular, we have $g_{VU} = g_{UV}^{-1}$ and $g_{UU} = \text{id}$.

9.2 Principal Bundles

Exercise 150 Corollary
Let $g_{UV}\colon U \cap V \to G$ be the transition functions of a trivial G-fibre bundle. Then there are continuous maps $f_U\colon U \to G$ on the cover sets U of the atlas so that $g_{UV} = f_U f_V^{-1}$ on $U \cap V$ holds.

Exercise 151 A Twisted Torus
On the edge of the cylinder $I \times S^1$, carry out the identification

$$(0, z) \sim (1, -z)$$

for all complex numbers $z \in S^1$. Show that the projection to S^1 on the first coordinate is trivial.

Exercise 152 Is So Typical
The discrete group G acts freely and properly on the topological space E. Then the base change from $p\colon E \to E/G$ along $p\colon E \to E/G$ is trivial.

Exercise 153 Withdrawn
Let $p\colon X \to B$ be a fibre bundle, and let

$$\begin{array}{ccc} X' & \longrightarrow & X \\ \downarrow p' & & \downarrow p \\ B' & \longrightarrow & B \end{array}$$

be a pullback. Then p' is a fibre bundle with the same typical fibre and structure group as p.

9.2 Principal Bundles

Every topological group G acts on itself by multiplication

$$G \longrightarrow \operatorname{Aut}(G), \quad g \longmapsto (f \mapsto gf).$$

Definition 9.2.1

The G-fibre bundles with typical fibre G and with these symmetries are called *G-principal bundles*. They are continuous maps $p\colon E \to B$ together with charts

$$\begin{array}{ccc} p^{-1}(U) & \xrightarrow{h_U} & U \times G \\ & {}_p\searrow \quad \swarrow_{\operatorname{pr}_U} & \\ & U & \end{array}$$

of an atlas so that the transition functions of p with respect to this atlas are given by continuous maps

$$g_{UV} \colon U \cap V \longrightarrow G.$$

The group G acts continuously on the product spaces $U \times G$ by

$$g(b, f) = (b, fg^{-1}).$$

The chart changes are G-equivariant because of

$$h_U h_V^{-1}(g(b, f)) = (b, g_{UV} fg^{-1}) = g(b, g_{UV} f) = g h_U h_V^{-1}(b, f).$$

So, they are G-homeomorphisms. This allows us to define a continuous G-action on E so that the charts are G-homeomorphisms: for $x \in E$ with $b = p(x) \in U$, let $h_U(x) = (b, g)$. Then the action is given by the formula

$$g'x = h_U^{-1}(b, gg'^{-1}).$$

Theorem 9.2.2
If $p\colon E \to B$ is a G-principal bundle, then the described action is free. The map induced by p from the orbit space E/G to B is a homeomorphism.

Proof. From $g'x = x$, it follows $gg'^{-1} = g$ and thus $g' = 1$. This proves the first assertion. Of course, we could have argued much faster and without formulas: the action is fibrewise and every fibre is G-homeomorphic to the free G-space G. This observation also shows that the map induced by p is bijective on the orbit space. To show that the inverse map is continuous, it is sufficient to restrict ourselves to open sets U in B over which the bundle is trivial. The map then has the form

$$U \cong U \times \{1\} \xrightarrow{\subseteq} U \times G \xrightarrow{h_U^{-1}} p^{-1}(U) \longrightarrow p^{-1}(U)/G \xrightarrow{\subseteq} E/G$$

and is thus continuous. □

Examples 9.2.3
We have already encountered principal bundles with discrete structure groups among the coverings, and we will again consider them separately in the following section. Another example of a principal bundle is the map

$$p \colon V_k(\mathbb{R}^n) \longrightarrow G_k(\mathbb{R}^n),$$

that assigns to a k-frame the subspace generated by it (see Sect. 5.2). It is an $O(k)$-principal bundle: let V be a k-dimensional subspace of $\mathbb{R}^n$, and let $V^\perp$ be its orthogonal complement and $p_V \colon \mathbb{R}^n \to$

9.2 Principal Bundles

V the orthogonal projection. Let U be the open set of all subspaces W with $W \cap V^\perp = 0$. Then the composite

$$A_W : W \subseteq \mathbb{R}^n \xrightarrow{p_V} V$$

is a linear isomorphism for all $W \in U$. Hence, the map A_W^{-1} sends an orthogonal basis of V to a basis of W. The Gram–Schmidt process even provides an orthogonal basis of W. Overall, this results in a local section

$$s : U \longrightarrow p^{-1}(U) \subseteq V_k(\mathbb{R}^n).$$

A trivialisation over U is thus given by

$$U \times O(k) \longrightarrow p^{-1}(U), \ (W, A) \mapsto s(W)A^{-1}.$$

After these examples of bundles, we clarify the relationships between these objects.

Definition 9.2.4

A *bundle map* between G-principal bundles p and p' is a G-map of the total spaces

$$f : E \longrightarrow E'.$$

Let p and p' be defined over the same base space B. Then f is called a bundle map over B if the map induced by f on the base space is the identity, i.e., if $p = p'f$ applies.

Obviously, principal bundles together with bundle maps form a category.

Theorem 9.2.5
Every principal bundle map over B is an isomorphism.

Proof. First, we consider the case when p and p' are product bundles $B \times G \to B$. For a bundle map f set

$$\alpha : B \longrightarrow G, \ \alpha(b) = \mathrm{pr}_G f(b, 1).$$

Then we have

$$f(b, g) = g^{-1} f(b, 1) = (b, \alpha(b)g).$$

So the map

$$(b, g) \mapsto (b, \alpha^{-1}(b)g)$$

is the inverse map and, hence, obviously continuous. For the general case, it follows that there is a cover by open sets of B over which the bundle map is invertible. Since f maps the fibres of p bijectively onto the fibres of p', the map f must also be bijective. The claim follows. □

Principal bundles are important because, among other things, we can construct a fibre bundle with typical fibre F and structure group G from a G-principal bundle $p \colon E \to B$ and an effective G-space F. This works as follows: the group G acts on E and F, consequently also on $E \times F$ by the product action

$$g(e, f) = (ge, gf).$$

Let $E \times_G F$ be the corresponding orbit space. The map p induces a map $E \times_G F \to B$. Charts for this map are obtained by those of p and the homeomorphisms

$$(U \times G) \times_G F \longrightarrow U \times F, \ ((b, x), f) \longmapsto (b, xf).$$

The transition functions of the fibre bundle $E \times_G F \to B$ are the same as those of p.

Definition 9.2.6

The bundle $E \times_G F \to B$ is called the *the fibre bundle associated to E* with typical fibre F.

This construction works just as well for G-spaces F on which G does not act effectively. This allows us to extend the definition of fibre bundles to such G-spaces. We quote the general definition, although we will only consider effective actions in the following. (Compare this definition with the definition in Sect. 9.1.)

Definition 9.2.7

A *G-fibre bundle* with typical fibre F is a map $p \colon X \to B$ together with a G-principal bundle $q \colon E \to B$ and an isomorphism h from $E \times_G F$ to X that maps the fibres of $E \times_G F$ onto the fibres of X:

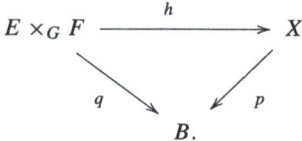

A *bundle map* of G-fibre bundles consists of a continuous map $f \colon X \to X'$ of the total spaces together with a G-principal bundle map $g \colon E \to E'$ that makes the diagram

$$\begin{array}{ccc} X & \xrightarrow{f} & X' \\ h \downarrow & & \downarrow h' \\ E \times_G F & \xrightarrow{g \times_G \mathrm{id}_F} & E' \times_G F \end{array}$$

commutative.

At this point, we can—as promised—provide an inverse to the equivalence from Theorem 8.5.14. It is given by the functor $M \mapsto U \times_G M$. Here, the universal covering $p \colon U \to B$ is considered as $\mathrm{Aut}(p)$-principal bundle, and $U \times_G M$ is the associated fibre bundle with typical fibre M, as described in Sect. 9.2.

Exercises

Exercise 154 Associated
The group $\mathbb{Z}/2$ acts on the circle S^1 through complex conjugation. The fibre bundle associated to $q \colon S^1 \to S^1$, $q(z) = z^2$ is the Klein bottle.

Exercise 155 Trivialised
A principal bundle is trivial if and only if it has a section.

Exercise 156 Multiplied
Show that the product $p \times p'$ of a G-principal bundle p and a G'-principal bundle p' is a $G \times G'$-principal bundle.

9.3 Principal Bundles with Discrete Structure Group

In the following, let X be a free G-space and G be discrete. Is the map $X \to X/G$ then a G-principal bundle? To answer this question, we recall the map

$$\theta \colon G \times X \longrightarrow X \times X, \quad (g, x) \longmapsto (x, gx).$$

It provides a continuous bijection θ' from $G \times X$ to the equivalence relation R corresponding to the surjection $X \to X/G$. The inverse map $R \to G \times X$ has the form

$$(x, x') \longmapsto (\varphi(x, x'), x)$$

for a map $\varphi\colon R \to G$. It assigns to each pair (x, x') of R the group element g that satisfies $x' = gx$. The group G acts properly on X if θ is proper. This is the case if and only if R is closed in $X \times X$ and θ' is a homeomorphism. This is the case if and only if R is closed in $X \times X$ and φ is continuous (see Sect. 5.3).

> **Theorem 9.3.1**
> *Let the discrete topological group G act freely on the space X. Then the following statements are equivalent.*
>
> (a) *The map $\varphi\colon R \to G$ is continuous.*
> (b) *The pre-image $\varphi^{-1}(1)$ is open.*
> (c) *Every point x of X has a neighbourhood U with $gU \cap U = \emptyset$ for all $g \neq 1$.*
> (d) *The map $p\colon X \to X/G$ is a G-principal bundle.*

Proof. (a) $\Rightarrow$ (b) Since G is discrete $\{1\}$ is open and hence so is the pre-image under the continuous map φ.

(b) $\Rightarrow$ (a) It suffices to show that for each group element g, the set $\varphi^{-1}(g)$ is open. But a homeomorphism of R is given by $(x, y) \mapsto (x, gy)$, which maps $\varphi^{-1}(1)$ to $\varphi^{-1}(g)$.

(b) $\Rightarrow$ (c) Let x be given. For the point (x, x) of $\varphi^{-1}(1)$ there is a neighbourhood of the form $(U \times U) \cap R$ with an open neighbourhood U of x in X that entirely lies in $\varphi^{-1}(1)$. If $U \cap gU$ is not empty, then there is a point u in U so that gu also lies in U. Then (u, gu) lies in $(U \times U) \cap R \subseteq \varphi^{-1}(1)$ and that means $g = 1$.

(c) $\Rightarrow$ (b) Let (x, y) be a point of $\varphi^{-1}(1)$, so $x = y$. If U is a neighbourhood of x as assumed then $(U \times U) \cap R$ is an open neighbourhood of (x, y) in R. It is entirely in $\varphi^{-1}(1)$; if $u' = gu$ with u and u' in U, then $U \cap gU \neq \emptyset$ follows, so $g = 1$.

(c) $\Rightarrow$ (d) If U is such a neighbourhood of x then $p(U)$ is open in X/G since p is open. The restriction of p to U provides a continuous, open surjection $U \to p(U)$. It is also injective because from $p(u) = p(u')$ it follows $u' = gu$ for an element g of G, so $U \cap gU \neq \emptyset$, so $g = 1$, and thus $u' = u$. This provides a homeomorphism $U \to p(U)$. Let $s\colon p(U) \to U$ be the inverse. Then

$$p(U) \times G \longrightarrow p^{-1}p(U),\ (b, g) \mapsto gs(b)$$

is a homeomorphism with inverse

$$p^{-1}p(U) \longrightarrow p(U) \times G,\ x \mapsto (p(x), \varphi(x, s(p(x)))^{-1}).$$

This provides a local trivialisation over $p(U)$.

(d) $\Rightarrow$ (c) The local trivialisations immediately provide such neighbourhoods. $\square$

9.3 Principal Bundles with Discrete Structure Group

We do not need the closedness of the relation R in proper actions to identify the quotient map as a principal bundle. Instead, it secures the Hausdorff property of the orbit space (see Sect. 5.3).

> **Corollary 9.3.2**
> Suppose the discrete group G acts freely and properly on the space X. Then the map $X \to X/G$ is a G-principal bundle, and the orbit space X/G is a Hausdorff space.

Every action of a compact Hausdorff group on a Hausdorff space is proper (see again Sect. 5.3).

> **Corollary 9.3.3**
> Let the finite group G act freely on the Hausdorff space X. Then the map $X \to X/G$ is a G-principal bundle, and the orbit space X/G is a Hausdorff space.

Example 9.3.4
Let $m \in \mathbb{Z}$ be an integer, and let p_1 and p_2 be two integers coprime to m. Then $\zeta_j = \exp(2\pi i p_j/m)$ for $j = 1, 2$ are two primitive m-th roots of unity and the action on the sphere $S^3 \subset \mathbb{C}^2$ given by

$$\mathbb{Z}/m \times S^3 \longrightarrow S^3, \ (k, (z_1, z_2)) \mapsto (\zeta_1^k z_1, \zeta_2^k z_2),$$

is free. The orbit space is the *lens space* $L(m; p_1, p_2)$. The projection

$$S^3 \longrightarrow L(m; p_1, p_2)$$

is a $\mathbb{Z}/m$-principal bundle.

Finally, it should be explained how the fundamental groups behave in principal bundles with discrete structure groups. This also is to be read in the light of covering theory.

> **Theorem 9.3.5**
> Let G be a discrete group, and let $p \colon E \to B$ be a G-principal bundle with path-connected total space E. For each point e in E, let α_e denote the bijection
>
> $$\alpha_e \colon G \longrightarrow M_p(p(e)), \ g \mapsto g^{-1}e$$

(continued)

> **Theorem 9.3.5** (continued)
> *between G and the fibre in e and*
>
> $$\beta_e \colon \pi_1(B, p(e)) \longrightarrow M_p(p(e)), \ [\gamma] \longmapsto [\gamma]e$$
>
> *be the map given by the fibre transport. Then*
>
> $$1 \longrightarrow \pi_1(E, e) \xrightarrow{p_*} \pi_1(B, p(e)) \xrightarrow{\alpha_e^{-1}\beta_e} G \longrightarrow 1$$
>
> *is an exact sequence of groups and group homomorphisms. In other words: p_* is injective, the map $\alpha_e^{-1}\beta_e$ is surjective and $\mathrm{Kern}(\alpha_e^{-1}\beta_e) = \mathrm{Image}(p_*)$. In particular, the subgroup $p_*\pi_1(E, e)$ is normal in $\pi_1(B, p(e))$.*

Proof. The important thing here is that the map is a homomorphism. Let γ be a loop at $p(e)$. If $\tilde{\gamma}$ is a lift of γ starting at e then $\gamma \mapsto g$ is equivalent to $\tilde{\gamma}(1) = g^{-1}e$. Let now γ' be another loop at $p(e)$ and $\tilde{\gamma}'$ a lift of it starting at e and ending at $(g')^{-1}e$. Then $g^{-1}\tilde{\gamma}'$ is a lift of γ' starting at $g^{-1}e$ and ending at $g^{-1}(g')^{-1}e = (g'g)^{-1}e$. Then is but $g^{-1}\tilde{\gamma}' \circ \tilde{\gamma}$ a lift of $\gamma' \circ \gamma$ starting at e and ending at $(g'g)^{-1}e$. It follows that $\gamma'\gamma \mapsto g'g$. The remaining statements follow directly from the previous theorem. Injectivity of p_* was shown there. The kernel of $\alpha_e^{-1}\beta_e$ is the stabiliser of e for the action given by fibre transport. Thus, it is equal to the image of p_* according to the previous theorem. Since E is path-connected, the group $\pi_1(B, p(e))$ acts transitively on the fibre according to the previous theorem. Hence β_e is surjective. □

We emphasise a special case of the preceding theorem:

> **Corollary 9.3.6**
> *Let $p \colon E \to B$ be a G-principal bundle of a discrete group G. If E is simply-connected, then every fundamental group of B is isomorphic to G.*

We should emphasise that, even in this situation, there is no distinguished isomorphism. To identify $\pi_1(B, b)$ with G, a point e above b had to be chosen, and the isomorphism determined by this choice depends on e.

Examples 9.3.7
The theorem and its corollary can be used to determine the fundamental groups of many spaces. For example, it again follows $\pi_1(S^1) \cong \mathbb{Z}$ and also $\pi_1(\mathbb{R}P^n) \cong \mathbb{Z}/2$ for $n \geq 2$.

Exercises

Exercise 157 Lensed
For the lens spaces (see Sect. 9.3), we have isomorphisms $\pi_1(L(m; p, q)) \cong \mathbb{Z}/m$.

9.4 Vector Bundles

The group $GL(n, \mathbb{R})$ acts canonically on the vector space $\mathbb{R}^n$.

Definition 9.4.1

The *real vector bundles* of *rank n* are the $GL(n, \mathbb{R})$-fibre bundles associated with these symmetries and with typical fibre $\mathbb{R}^n$. They are locally trivial maps $p \colon V \to B$ together with charts

$$\begin{array}{ccc} p^{-1}(U) & \xrightarrow{h_U} & U \times \mathbb{R}^n \\ & \searrow p \quad \swarrow \text{pr}_U & \\ & U & \end{array}$$

of a (maximal) atlas whose transition functions are given by linear isomorphisms of $\mathbb{R}^n$. We can use the vector space structure on $\mathbb{R}^n$ to define the vector space structure on each fibre. Similarly, we define complex vector bundles with $\mathbb{C}$ instead of $\mathbb{R}$.

Examples 9.4.2
Important examples of vector bundles are given by the tangent bundles of smooth manifolds. Thus, the projection of

$$\{(b, x) \in S^n \times \mathbb{R}^{n+1} \mid \langle b, x \rangle = 0\}$$

onto the first factor is the tangent bundle of the *n*-sphere. The fibre over a point is the orthogonal complement of the line passing through this point. A chart over $U_j = \{b \in S^n \mid b_j \neq 0\}$ is given by the inverse of

$$U_j \times \mathbb{R}^n \longrightarrow p^{-1}(U_j), \ (b, y) \mapsto (b, v_b(y_1, \ldots, y_j, 0, y_{j+1}, \ldots, y_n)),$$

where $v_b(x) = x - \langle b, x \rangle b$ is the normal component of x to b. For fixed b, this map is linear and maps the standard basis to a basis of the tangent space at b. There are also other important examples of vector bundles: the projection map onto $G_k(\mathbb{R}^n)$ with source

$$\{(V, v) \in G_k(\mathbb{R}^n) \times \mathbb{R}^k \mid v \in V\}$$

is called the *tautological bundle* because the fibre over V is the vector space V itself. It is isomorphic to the bundle that associates the fibre $\mathbb{R}^k$ to the $O(k)$-principal bundle $V_k(\mathbb{R}^n)$ (see the example in Sect. 9.2) and therefore is locally trivial. For $k = 1$ and $n = 1$, it contains the interval bundle, i.e. the set of all (V, v) with $|v| \leqslant 1$. This bundle is the Möbius strip.

The study of vector bundles over a space B is 'linear algebra over B', just as vector bundles over a point are the same as vector spaces. Many constructions for vector spaces also exist for vector bundles: direct sum, tensor product, duality and general Hom-bundles. (The external powers of vector bundles are especially important for the Cartan calculus of analysis on manifolds.) In this view, new phenomena arise that are not visible over $B = \star$. For example, every vector space is orientable, but not every vector bundle. Similarly, every finite-dimensional vector space is isomorphic to its dual space, but not every finite rank vector bundle is isomorphic to its dual. Some constructions are, therefore, only locally feasible but not globally.

Finally, we should note that when examining vector bundles, it is advisable not to predetermine the dimension of the fibres. For example, if the base is a sum of topological spaces, the fibre dimension can vary on the summands, thus distinguishing these summands. A vector bundle in this sense over a space B is a continuous map $p: V \to B$ with the structure of a vector space on each fibre so that every point b from B has a neighbourhood U over which there is a trivialisation $p^{-1}(U) \cong U \times p^{-1}(b)$ that is an isomorphism of vector spaces in each fibre. The transition functions are then automatically given by isomorphisms of vector spaces.

Supplements

Classification of Vector Bundles The tautological bundles play a decisive role in the classification of vector bundles. Only the following result is mentioned here: if $p: X \to B$ is a real vector bundle of dimension k over a compact Hausdorff space B, then for sufficiently large n, there always exists a continuous map $f: B \to G_k(\mathbb{R}^n)$ with the property that the bundle that is created by pulling back the tautological bundle along f is isomorphic to p. The map f can be replaced by any homotopic map and is also uniquely determined up to homotopy. In this way, a bijection between isomorphism classes of vector bundles and homotopy classes of maps into the Grassmann manifolds is obtained (for details, see for example [Hus94]).

Flat Vector Bundles When interpreting vector bundles as $GL(n, \mathbb{R})$-fibre bundles with typical fibre $\mathbb{R}^n$ it is crucial to equip the matrix group $GL(n, \mathbb{R})$ with the usual topology. It could also have been equipped with the discrete topology. This would have led to the concept of *flat* vector bundles of rank n. Every flat vector bundle is a vector bundle because the identity $GL(n, \mathbb{R}) \to GL(n, \mathbb{R})$ is continuous if the source has the discrete and the target the classical topology. However, there are vector bundles that are not flat.

Exercises

Exercise 158 Tangent to Lines
The tangent bundle of $\mathbb{R}P^n$ is the quotient bundle of the tangent bundle of S^n where (b, x) is identified with $(-b, -x)$. Construct an atlas.

Exercise 159 Bases of Vector Bundles
Show that a k-dimensional vector bundle $V \to B$ is trivial if and only if there are k sections $v_1, v_2, \ldots, v_k$ whose values $v_1(b), v_2(b), \ldots, v_k(b)$ form a basis for each b in the fibre over b. Only trivial vector bundles, therefore, have bases.

9.5 Fibrations

In this section, we introduce the concept of a fibration between topological spaces. These are maps that have lifting properties regarding homotopies. Unlike the coverings, we only require the existence (but not the uniqueness) of liftings. In addition, the exact sequence from Theorem 9.3.5 is generalised in this context.

Definition 9.5.1

Let $p\colon X \to Y$ be a continuous map between topological spaces. Then p is called a *fibration* (or more precisely, a *Serre fibration*), if p has the following lifting property: in every commutative diagram with solid arrows of the form

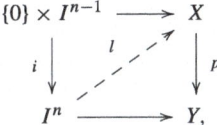

there exists a lifting l that makes the two triangles commutative.

There is an automorphism of I^n that maps the subspace $\{0\} \times I^{n-1} \cup I \times \partial I^{n-1}$ onto the subspace $\{0\} \times I^{n-1}$. Such is shown in Fig. 9.4.

With this, we obtain:

▶ **Remark 9.5.2** A map is a fibration if and only if it has the lifting property for all inclusions of the form $\{0\} \times I^{n-1} \cup I \times \partial I^{n-1} \subseteq \{0\} \times I^{n-1}$.

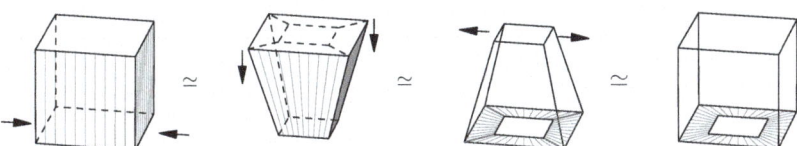

Fig. 9.4 A relative automorphism

Examples 9.5.3
Every covering is a fibration according to Theorem 8.2.2. In that case, the liftings even fulfil a uniqueness statement. Every trivial map $\mathrm{pr}_B \colon B \times F \to B$ is a fibration because a lifting $l \colon I^n \to B \times F$ in the first factor is already prescribed and because a map I^{n-1} can always be extended to I^n using the projection.

We can obtain further examples by the following theorem.

Theorem 9.5.4
Every fibre bundle $p \colon X \to B$ is a fibration.

Proof. It suffices to verify the following statement: if $(U_j \mid j \in J)$ is an open cover of B such that all restrictions $p^{-1}U_j \to U_j$ are fibrations, then this also applies to p. To see this, choose for a lifting problem a subdivision of the cube I^{n-1} into subcubes $(W_i \mid i \in I)$ and a decomposition of the interval $t_0 = 0 < t_1 < \cdots < t_m = 1$ with the property that the homotopy maps each individual subcube $[t_k, t_{k+1}] \times W_i$ already into a U_j (see Lebesgue number, Sect. 4.1). Each subcube W_i is itself (for each d) divided into its d-dimensional sides W_i^k and arranged according to dimension. Now, we successively solve the lifting problem

$$\begin{array}{ccc} \{t_l\} \times W_i^k \cup [t_l, t_{l+1}] \times \partial W_i^k & \longrightarrow & p^{-1}U_j \\ \downarrow & & \downarrow \\ [t_l, t_{l+1}] \times W_i^k & \longrightarrow & U_j. \end{array}$$

This is possible according to the remark. □

Some fibrations do not come from fibre bundles:

Theorem 9.5.5
For each k and every space X, the map

$$i^* \colon \mathrm{Hom}(I^k, X) \longrightarrow \mathrm{Hom}(\partial I^k, X)$$

induced by the inclusion $i \colon \partial I^k \to I^k$ between the mapping spaces is a fibration.

9.5 Fibrations

Fig. 9.5 A retraction

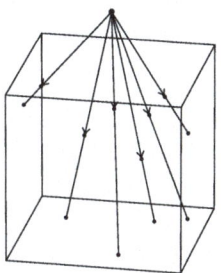

Proof. The exponential law shows that the lifting problem

$$\begin{array}{ccc} \{0\} \times I^{n-1} & \longrightarrow & \mathrm{Hom}(I^k, X) \\ i \downarrow & \nearrow^{l} & \downarrow \\ I^n & \longrightarrow & \mathrm{Hom}(\partial I^k, X) \end{array}$$

is equivalent to the extension problem

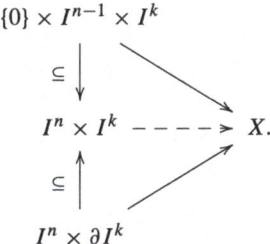

This is always solvable because the subspace $\{0\} \times I^{n-1} \times I^k \cup I^n \times \partial I^k$ is a retract of $I^n \times I^k$. A retraction is indicated in Fig. 9.5. □

Example 9.5.6
The case $n = 1$ corresponds to the map

$$(s, t) \colon P(X) = \mathrm{Hom}(I, X) \longrightarrow \mathrm{Hom}(\partial I, X) = X \times X$$

from Sect. 6.1 and is called the *path space fibration*. It assigns to a path in X its start and endpoint.

Once a fibration has been found, it is easy to construct more.

Theorem 9.5.7
Fibrations are preserved under base change.

Proof. Let

$$\begin{array}{ccc} X' & \longrightarrow & X \\ p' \downarrow & & \downarrow p \\ Y' & \longrightarrow & Y \end{array}$$

be a pullback and p a fibration. Also, let there be given a lifting problem of the form

$$\begin{array}{ccc} \{0\} \times I^{n-1} & \longrightarrow & X' \\ \downarrow & & \downarrow p' \\ I^n & \longrightarrow & Y' \end{array}.$$

Then the two squares can be connected horizontally, and a lift $l: I^n \to X$ is obtained. Using the universal property of the pullback, we find that l and the morphism from I^n to Y' define the desired lift $l': I^n \to X'$. □

> **Corollary 9.5.8**
> For $x \in X$, let $P(X, x)$ be the space of paths in X that start in x. Then the map
>
> $$t: P(X, x) \longrightarrow X$$
>
> that assigns the endpoint to a path is a fibration. The fibre over x is $\Omega(X, x)$.

Proof. The diagram

$$\begin{array}{ccc} P(X, x) & \longrightarrow & P(X) \\ \downarrow & & \downarrow (s,t) \\ X & \xrightarrow{(x,\mathrm{id})} & X \times X \end{array}$$

is a pullback. □

The following observation generalises the exact sequence of Theorem 9.3.5.

9.5 Fibrations

Theorem 9.5.9 (Exact Sequence of a Fibration)
Let $p: X \to Y$ be a fibration and denote by $i: F = p^{-1}(y) \subseteq X$ the inclusion of the fibre over a point $y \in Y$. Then for each $x \in F$ the sequence

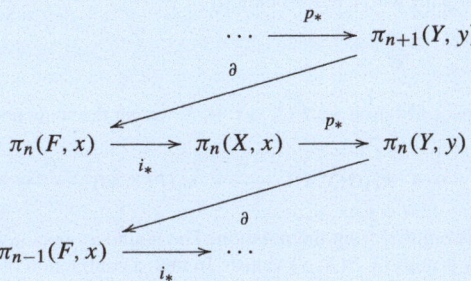

is exact. The maps $\partial: \pi_n(Y, y) \to \pi_{n-1}(F, x)$ are given by $\partial[\gamma] = [t \mapsto \tilde{\gamma}(1, t)]$ where $\tilde{\gamma}$ is a lift of the map $\gamma: I^n \to Y$ to X that maps the subspace $\{0\} \times I^{n-1} \cup I \times \partial I^{n-1}$ to the base point. The sequence is natural with respect to base change.

Proof. First, we must show that the given boundary map is well-defined. Suppose γ' is another representative of a class in $\pi_n(Y, y)$, and $\tilde{\gamma}'$ is a lift. Then the restrictions to the back side should be bounded homotopic in F. Such a homotopy is obtained by lifting a homotopy h between γ and γ' where all sides except the back are prescribed (see Fig. 9.6).

The compositions of two maps each are trivial. This directly results from their definition. The inclusion of Kern(i_*) $\subseteq$ Im(∂) comes from the projection of a null-homotopy in X of a representative of $\pi_n(F, x)$ to Y. The inclusion of Kern(p_*) $\subseteq$ Im(i_*) can be seen when a null-homotopy in Y to X is lifted to the given start point. Finally, the inclusion Kern(∂) $\subseteq$ Im(p_*) remains. It is obtained by composing a lift of the image to Y with a null-homotopy in F (see Fig. 9.7). □

Fig. 9.6 A lifted homotopy

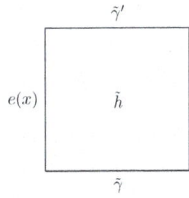

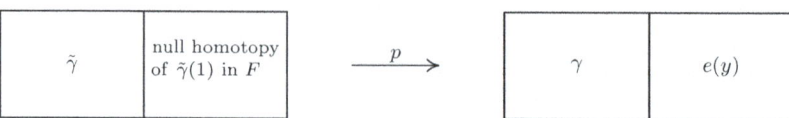

Fig. 9.7 A composite of a lift with a null-homotopy

Example 9.5.10
In the case of the path space fibration $p\colon P(X, x) \to X$ we get the sequence

$$\cdots \xrightarrow{\ \partial\ } \pi_{n+1}(X) \xrightarrow{\ \partial\ } \pi_n(\Omega(X, x)) \xrightarrow{\ i_*\ } \pi_n(P(X, x)) \xrightarrow{\ p_*\ } \pi_n(X) \xrightarrow{\ \partial\ } \cdots,$$

where the base point was omitted from the notation. The boundary operator ∂ is an isomorphism. Therefore, the homotopy groups of $P(X, x)$ vanish. In fact, a contraction of $P(X, x)$ can easily be written down:

$$I \times P(X, x) \longrightarrow P(X, x), \quad (w, t) \mapsto (s \mapsto w(st)).$$

Supplements

Cofibrations We can ask which maps can be lifted with given initial values besides homotopies. This quickly leads to the concept of *cofibrations*. These are those maps $i\colon A \to B$ for which the lifting problem

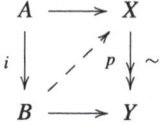

is solvable for all acyclic fibrations p. In this context, a map is called *acyclic* if it provides isomorphisms on all homotopy groups for all choices of base points. If B is formed by attaching cells to A, then B has this property. If i is also acyclic, then i has the lifting property for all fibrations.

Homotopy Groups of Spheres We have already determined π_0 and π_1 of the spheres S^n. But how about the higher homotopy groups $\pi_m S^n$ for $m \geqslant 2$? The long exact sequence for the exponential fibration shows $\pi_m S^1 = 0$, for all $m \geqslant 2$. So we only need to worry about $n \geqslant 2$. From now on, let thus $m, n \geqslant 2$. It can be shown that $\pi_m S^n = 0$ holds if $m < n$ is fulfilled. A key idea is the following: we show that every continuous map can be homotoped to a map that lies in a particularly nice class of maps; and then we show that all these nice maps are null-homotopic. It can also be shown that $\pi_m S^n$ is isomorphic to $\mathbb{Z}$ if $m = n$ holds. Isomorphisms are given by the suspension: let $f\colon S^m \to S^n$ be continuous. Then their *suspension*

$$E(f)\colon S^{m+1} \to S^{n+1},$$

9.5 Fibrations

is defined by

$$(t, v) \longmapsto \begin{cases} (t, \sqrt{1-t^2} f(v/\|v\|)) & v \neq 0, \\ (t, 0) & v = 0. \end{cases}$$

The suspension theorem of Freudenthal (see [Hat02]) states that the homomorphisms given by suspension

$$\pi_{m+k} S^{n+k} \to \pi_{m+k+1} S^{n+k+1} \to \pi_{m+k+2} S^{n+k+2} \to \cdots$$

are bijective if $k \geqslant m - 2n + 2$ holds. The common value then only depends on the difference $m - n$ and is often denoted by π_{m-n}^{st}. These are the *stable homotopy groups* of the spheres. We may hope that these are easier to calculate than the individual $\pi_m S^n$. This is true, but that does not necessarily mean it is simple. The following table gives the isomorphism type of some of these groups.

k	0	1	2	3	4	5	6	7	8
π_k^{st}	$\mathbb{Z}$	$\mathbb{Z}/2$	$\mathbb{Z}/2$	$\mathbb{Z}/24$	0	0	$\mathbb{Z}/2$	$\mathbb{Z}/240$	$\mathbb{Z}/2 \oplus \mathbb{Z}/2$

Stable homotopy theory deals, among other things, with finding an order in this structure.

Exercises

Exercise 160 The Hopf Map
Use the fibre bundle $S^3 \to \mathbb{CP}^1 \cong S^2$ with fibre S^1 to construct an isomorphism

$$\pi_3(S^2) \cong \pi_3(S^3).$$

Sheaves

10

The sheaf concept is part of the basic vocabulary of modern geometry and topology. It serves to formalise the transition from the local to the global. However, sheaf theory is much more general; it can even be used to study the foundations of logic. These statements already indicate the generality of the theory. Therefore, we will only try to introduce the basic ideas here leisurely, without claiming to deploy the category-theoretical machine properly. We will naturally not get as far as, for example, the book [MM94], which is well suited for further education in this direction.

10.1 Presheaves and Sheaves

Sheaves allow us to neglect the points of a topological space and focus entirely on its open subsets, their inclusions, and their cover properties. We formalise this idea as follows.

Definition 10.1.1

For a topological space X, let $\mathcal{U}(X)$ be the subcategory of the category of topological spaces and continuous maps that consists of the open subsets U of X and the inclusions between them. This is the *site* associated with X.

We can define sites much more generally and then consider sheaves on these. However, we will not do this here to avoid getting sidetracked even before the journey into the land of sheaves has begun properly.

Definition 10.1.2

Let X be a topological space. The category $\mathbf{Pr}(X)$ of the *presheaves* on X is the category of functors from $\mathcal{U}(X)^{\mathrm{op}}$ to the category **Sets** of sets (see Sect. 1.4 for the opposite category).

We now want to expand this definition briefly. A presheaf M assigns to each open subset U of X a set $M(U)$. The elements of $M(U)$ are often called the *sections* of M over U. If $U \subseteq V$, then the presheaf M assigns to this inclusion a map $M(V) \to M(U)$. The image of a section s over V under this map is called *restriction* of s to U and is usually simply denoted by $s|U$.

Examples 10.1.3
For every space X, there is the presheaf C of continuous $\mathbb{R}$-valued functions on X. Here, the set $C(U)$ consists of the continuous functions $U \to \mathbb{R}$. The restriction maps are given by restricting functions. The following class of examples is of paramount importance. We could almost say, and this will be specified below in Theorem 10.2.6, that there are no other examples. Let $p: E \to X$ be a continuous map. This defines a sheaf, which is denoted by hE, by

$$hE(U) = \{s: U \to E \mid ps(u) = u \text{ for all } u\}$$

on objects, as well as by restriction on morphisms. Particularly interesting from a systematic point of view are the maps $p: E \to X$ of the form $U \subseteq X$; these provide the *representable presheaves*. If M is any set, we can consider the *constant presheaf* $M(U) = M$. Its restriction maps are given by id_M.

The morphisms between presheaves are given by the natural transformations between them. A morphism $\varphi: M \to N$ thus consists of a family of maps $\varphi(U): M(U) \to N(U)$, one for each U, so that for each inclusion $U \subseteq V$ the diagram

$$\begin{array}{ccc} M(U) & \xrightarrow{\varphi(U)} & N(U) \\ \uparrow & & \uparrow \\ M(V) & \xrightarrow{\varphi(V)} & N(V) \end{array}$$

commutes.

The concept of a presheaf on X is too general to store topological properties of X. It only uses the inclusions of the open subsets from X, but not any information on when and how these overlap. The sheaf concept takes this data into account.

Definition 10.1.4

Let X be a topological space. The category $\mathbf{Sh}(X)$ of *sheaves* on X is the full subcategory of the category of presheaves M on X with the following property: If $(U_j \subseteq X \mid j \in J)$ is a cover of an open subset $U \subseteq X$, then there exists for

10.1 Presheaves and Sheaves

each family $(s_j \in M(U_j) \mid j \in J)$ of sections with

$$s_j|(U_j \cap U_k) = s_k|(U_j \cap U_k)$$

for all j and k a unique section $s \in M(U)$ with $s_j = s|U_j$ for all j.

A sheaf is, therefore, a presheaf whose sections are determined locally and can also be specified locally.

Examples 10.1.5
The presheaf hE of sections of a continuous map $E \to X$ is always a sheaf on X. We obtain thus a functor h from the category **Top**(X) of spaces over X into the category **Sh**(X). The sheaf of sections of the projection $\mathrm{pr}_X \colon X \times \mathbb{R} \to X$ is isomorphic to the presheaf C of continuous $\mathbb{R}$-valued functions on X. In particular, the presheaf C is a sheaf. The representable presheaves are, therefore, also sheaves.

A constant presheaf does not have to be a sheaf because an assignment that is locally constant does not have to be globally constant. 'Constancy' is, therefore, a property that cannot be checked locally; 'continuity', on the other hand, can. If S is a set, we can define the sheaf of *locally constant functions with values in S*; this is simply the sheaf of continuous maps to S, when S is equipped with the discrete topology. If M is a sheaf on X that has at least one section, then $M(\emptyset)$ is a singleton. Any two restrictions of sections agree on the empty subset. Products of presheaves can be formed objectwise:

$$(\prod_\lambda M_\lambda)(U) = \prod_\lambda (M(U_\lambda)).$$

Here, the product of sheaves is again a sheaf.

Exercises

Exercise 161 Sheaf Pullback
The pullback of a diagram of presheaves is formed objectwise and retains the sheaf property. What about pushouts?

Exercise 162 Generalised Sets
The category **Sh**$(\star)$ of sheaves on the one-point space $\star$ is isomorphic to the category of sets. In this sense, sheaves are generalised sets.

Exercise 163 Restrictions also from Sheaves
If M is a sheaf on X and $U \subseteq X$ an open subset, then we can define, by restriction of the functor M from $\mathcal{U}(X)$ to the open subsets of U, a sheaf $M|U$ on U. (From the current standpoint, it is not foreseeable how this construction can be generalised to arbitrary subsets of X—or even continuous maps from any space to X. These will be the pullbacks of sheaves, defined in Sect. 10.3.)

Exercise 164 The Morphism Sheaf
Let M and N be sheaves on X. Show that the presheaf

$$U \longmapsto \mathrm{Mor}_{\mathsf{Sh}(X)}(M|U, N|U)$$

is a sheaf.

Exercise 165 Gluing Sheaf
Suppose that $(U_j \mid j \in J)$ is an open cover of X and M_j are sheaves on U_j. Let

$$\varphi_{ij} \colon M_i|(U_i \cap U_j) \longrightarrow M_j|(U_i \cap U_j)$$

be morphisms with $\varphi_{ii} = \mathrm{id}$ and $\varphi_{ik} = \varphi_{jk}\varphi_{ij}$ on all $U_i \cap U_j \cap U_k$. Then there is a unique sheaf on X and isomorphisms $\psi_i \colon M|U_i \to M_i$ with $\psi_j = \varphi_{ij}\psi_i$ on all $U_i \cap U_j$.

10.2 Stalks and Étale Spaces

In this section, we will assign to each sheaf over X a topological space over X that will have the same local sections. For this, we first deal with the fibres for this construction.

Definition 10.2.1

Let M be a presheaf on a topological space X. For each point x from X, let $\mathcal{U}(X, x)$ be the full subcategory of the category $\mathcal{U}(X)$ of open sets of X that contain x. By restricting M, we obtain a functor from $\mathcal{U}(X, x)^{\mathrm{op}}$ to the category of sets. We define

$$M_x = \coprod_{U \in \mathcal{U}(X,x)} M(U)/\sim,$$

where $\sim$ is the following equivalence relation. Two sections s on U and t on V are equivalent if and only if there is an open neighbourhood W of x that is contained in U and in V and on which $s|W = t|W$ holds. The equivalence classes $[S, U]$ are called *germs*, and M_x is called the *stalk* (or sometimes also the *fibre*) of M over x.

The germ $[S, U]$ of a section s on U thus remembers from s only the behaviour in arbitrarily small neighbourhoods of x.

Examples 10.2.2
The stalks of the sheaf L of locally constant maps on X with values in a set S can be identified with S via the evaluation. The map

$$L_x \longrightarrow S, \quad [f, U] \longmapsto f(x)$$

10.2 Stalks and Étale Spaces

is a bijection. For the germs of real-valued continuous functions, the function value is also defined but it does not always provide a bijection.

Example 10.2.3
If y is a point in Y and M any set, then a sheaf on Y is given by

$$U \longmapsto \begin{cases} M & y \in U, \\ \star & y \notin U. \end{cases}$$

This sheaf is the *skyscraper sheaf* on Y (with stalk M in y). It should be noted here that not only can the stalk at y be identified with M; this is also true for the stalks at the points in every open neighbourhood of y. For Hausdorff spaces, this is, of course, only y itself. All other stalks are singletons.

We will now assign a topological space to each presheaf M on X. We will proceed in a similar way as in the construction of a covering from a $\Pi(X)$-set in Sect. 8.4. We therefore consider the disjoint union

$$|M| = \coprod_{x \in X} M_x$$

of all stalks. Then there is a canonical map $p\colon |M| \to X$; it maps the stalk M_x to x. We want to define a topology on $|M|$ such that this map is continuous. For this, we consider the sections $s \in M(U)$ of M. Each such section defines a map

$$U \longrightarrow |M|, \quad u \longmapsto [S, U],$$

which is also denoted by s. We have $ps(u) = u$ for all u in U. We give $|M|$ the topology co-induced by all sections s. With respect to this topology, the maps $p\colon |M| \to X$ and $s\colon U \longrightarrow |M|$ are continuous.

Definition 10.2.4

The space $|M|$ is called the *étale space* of M (after the French adjective *étale*).

Theorem 10.2.5
The map $p\colon |M| \to X$ is a local homeomorphism.

Proof. A point of $|M|$ is a germ of a section s of M over a subset $U \subseteq X$ at a point. The set $s(U)$ is an open neighbourhood of this point in $|M|$, because if (t, V) is a section over V and w is a point in $t^{-1}(s(U))$, then w is in $U \cap V$, and because $s(w) = t(w)$, we know that s and t agree in a small neighbourhood $W \subseteq U \cap V$ of w. Therefore, the set W is in $t^{-1}(s(U))$ and $s(U)$ is open. The map p maps $s(U)$ homeomorphically onto U. □

The construction just described is compatible with sheaf morphisms. Thus, for every topological space X, we obtain a functor

$$|?|: \mathbf{Pr}(X) \longrightarrow \mathbf{Et}(X)$$

into the full subcategory $\mathbf{Et}(X)$ of the category $\mathbf{Top}(X)$ of topological spaces over X whose structure map is a local homeomorphism. Overall, we have now defined a cycle

$$\begin{array}{ccc} \mathbf{Sh}(X) & \xrightarrow{\subseteq} & \mathbf{Pr}(X) \\ {\scriptstyle h}\uparrow & & \downarrow {\scriptstyle |?|} \\ \mathbf{Top}(X) & \xleftarrow{\supseteq} & \mathbf{Et}(X) \end{array}$$

of functors. We use these now to complete the basics of sheaf theory. The following result is the main theorem.

> **Theorem 10.2.6 (Main Theorem About Étale Spaces and Sheaves)**
> Let X be a topological space. The functor
>
> $$|?|: \mathbf{Sh}(X) \longrightarrow \mathbf{Et}(X)$$
>
> is left adjoint to the functor h,
>
> $$\mathrm{Mor}_{\mathbf{Et}(X)}(|M|, E) \cong \mathrm{Mor}_{\mathbf{Sh}(X)}(M, hE),$$
>
> and an equivalence of categories (with inverse h).

Proof. If M is a presheaf on X, we assign to a section s of M over U the continuous map $U \to |M|$ that—as described in Sect. 10.2—is defined by this section. This provides a morphism

$$\eta: M \longrightarrow h|M|, \quad s \mapsto [S, U].$$

If M is a sheaf, this is an isomorphism. To see this, let $s: U \to |M|$ be a continuous section over U. For $x \in U$ choose (s_x, U_x) with $s(x) = [S_x, U_x]$. If $z \in U_x \cap U_y$, then

$$[S_y, U_y] = s(z) = [S_x, U_x]$$

10.2 Stalks and Étale Spaces

and thus $s_x|W = s_y|W$ for an open neighbourhood W of z. Because M is a sheaf, the sections s_x and s_y agree on $U_x \cap U_y$, and together define a unique section $s \in M(U)$. Conversely, if E is a space over X, we can, for each germ (in x) of a continuous section s of E over U, evaluate this section in x and thus obtain a point in E. This provides a map

$$\varepsilon : |hE| \longrightarrow E, \ [S, U]_x \mapsto s(x)$$

over X. It is continuous because s is continuous. If E is also a local homeomorphism over X. It is even a homeomorphism because then, locally, the continuous inverse map can easily be written down. This gives the equivalence of the categories. The adjunction bijections are given by

$$(f : |M| \longrightarrow E) \mapsto (\ M \xrightarrow{\varphi} h|M| \xrightarrow{h(f)} hE\ )$$

in one direction and by

$$(\varphi : M \longrightarrow hE) \mapsto (\ |M| \xrightarrow{|\varphi|} |hE| \xrightarrow{\varepsilon} E\ )$$

in the other direction. □

The theorem states, among other things, that the concept of a sheaf is, in principle, superfluous because it can be replaced at any time by the equivalent concept of the étale space. While this may be logically correct, it is psychologically advantageous to be familiar with the sheaf perspective. The conceptual world of sheaf theory leads to ideas that might not be so obvious from the standpoint of the étale spaces.

Exercises

Exercise 166 Ask Taylor
For the sheaf of analytical functions on an open subset of $\mathbb{R}$, the stalks are given by the set of all convergent power series.

Exercise 167 One and Co-one
Let

$$\mathcal{C} \underset{R}{\overset{L}{\rightleftarrows}} \mathcal{D}$$

be adjoint functors, i.e., there is a natural bijection

$$\mathrm{Mor}_{\mathcal{D}}(L(X), Y) \cong \mathrm{Mor}_{\mathcal{C}}(X, R(Y)).$$

Then there is a *unit* natural transformation $\eta \colon \mathrm{id}_{\mathcal{C}} \to RL$ and another natural transformation $\varepsilon \colon RL \to \mathrm{id}_{\mathcal{D}}$ (a *co-unit*) for which the compositions

$$R \xrightarrow{\eta R} RLR \xrightarrow{R\varepsilon} R \;, \quad L \xrightarrow{L\eta} LRL \xrightarrow{\varepsilon L} L$$

are identities. Conversely, a unit η and a co-unit ε, for which these compositions are identities, always define an adjunction between L and R.

10.3 Sheafification and Pullbacks

The results of the previous section can be used to turn a pre-sheaf into a sheaf. This is obviously useful. Whenever we encounter a pre-sheaf where we cannot already tell from a distance that it is a sheaf, we sheafify it, just to be safe. The construction is particularly helpful when transferring geometric morphisms into the world of sheaves.

Definition 10.3.1

Let X be a topological space. If M is a pre-sheaf on X, then $\mathrm{h}|M|$ is a sheaf on X. This is called the *sheafification* of M.

Theorem 10.3.2
The sheafification functor $\mathbf{Pr}(X) \to \mathbf{Sh}(X)$ *is left adjoint to the forgetful functor:*

$$\mathrm{Mor}_{\mathbf{Sh}(X)}(\mathrm{h}|M|, N) \cong \mathrm{Mor}_{\mathbf{Sh}(X)}(M, N).$$

Proof. If $M \to N$ is a morphism of sheaves, then sheafification of it provides a morphism $\mathrm{h}|M| \to \mathrm{h}|N|$. If N is already a sheaf, then the target $\mathrm{h}|N|$ is isomorphic to N, which by composition provides a morphism $\mathrm{h}|M| \to N$.

$$\begin{array}{ccc} M & \longrightarrow & N \\ \downarrow & & \downarrow \cong \\ \mathrm{h}|M| & \longrightarrow & \mathrm{h}|N| \end{array}$$

This assignment is bijective; the inverse is given by composition with the universal morphism $M \to \mathrm{h}|M|$. □

We may think of the sheafification of a presheaf M as a 'free sheaf' on M, if that helps the intuition.

10.3 Sheafification and Pullbacks

Example 10.3.3
The sheafification of a presheaf of constant functions is the corresponding sheaf of locally constant functions.

If $f: X \to Y$ is a continuous map, then we obtain by taking pre-images (think: 'pullbacks') a functor

$$f^{-1}: \mathcal{U}(Y) \to \mathcal{U}(X), \quad V \mapsto f^{-1}V$$

in the opposite direction. By composition, we then obtain a functor

$$f_*: \mathbf{Pr}(X) \longrightarrow \mathbf{Pr}(Y)$$

that goes back in the original direction.

▶ **Remark 10.3.4** The functor f_* maps sheaves to sheaves.

It is not hard to see that a collection of sections

$$s_j \in f_*(M)(V_j) = M(f^{-1}(V_j))$$

that agree on intersections define a unique section on the union because this is the case for M.

Examples 10.3.5
We shall discuss two distinguished examples: when $X = \{x\}$ or when $Y = \{y\}$. If $Y = \{y\}$, then f_*M is a sheaf on the one-point space, thus a set, namely the set $M(X)$ of *global sections* of M. If, on the other hand, we take $X = \{x\}$, then the sheaf M on it is simply a set. The sheaf f_*M is then given by

$$(f_*M)(V) = \begin{cases} M & f(x) \in V \\ \star & f(x) \notin V \end{cases}$$

given, thus by the skyscraper sheaf on Y with stalk M in $f(x)$.

If M is a sheaf on Y, then its *pullback* is defined by

$$f^*M = \mathrm{h}(X \times_Y |M|)$$

So, we simply consider the étale space belonging to M over Y and then the sheaf belonging to its pullback.

> **Theorem 10.3.6**
> The pullback functor f^* is left adjoint to f_*:
> $$\mathrm{Mor}_{\mathsf{Sh}(X)}(f^*M, N) \cong \mathrm{Mor}_{\mathsf{Sh}(Y)}(M, f_*N).$$

Proof. If s is a section of N over $V \subseteq Y$, it can be considered as a continuous section of $|N|$ over V. The universal property of the pullback provides a section of the pullback $X \times_Y |N|$ over the pre-image $f^{-1}V$, thus a section of $f_* f^* N$ over V. This defines a morphism

$$\eta\colon N \longrightarrow f_* f^* N.$$

Conversely, if s is a section of $f^* f_* M$ over $U \subseteq X$, i.e., a continuous map from U to $X \times_Y |f_*M|$, the second component thereof provides a section of M over U. This defines a morphism

$$\varepsilon\colon f^* f_* M \longrightarrow M.$$

The adjunction now follows as in the proof of Theorem 10.2.6, or we use one of the exercises from Sect. 10.2. □

Examples 10.3.7
The same examples as before apply: $X = \{x\}$ or $Y = \{y\}$. If we start with $Y = \{y\}$, then f^*N is the sheaf of locally constant functions on X with values in N. However, if $X = \{x\}$, then f^*N is a sheaf on the one-point space, thus a set, the stalk of N in $f(x)$.

Supplement

Abelian Sheaves and So On and So Forth So far, we have only considered sheaves of sets. Of course, we can also consider 'structured' sheaves. For example, an *abelian sheaf* on $\mathcal{U}(X)$ is a sheaf M on $\mathcal{U}(X)$ together with a factorisation

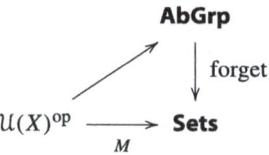

through the forgetful functor. This means that all sets $M(U)$ are equipped with the structure of an abelian group so that the restriction maps are group homomorphisms. Similarly, we can then also define *ring sheaves* and *module sheaves* and generalise all of the usual algebra in this context.

Exercises

Exercise 168 Restrictions Are Pullbacks
If $i: U \to X$ is the inclusion of an open subset, then $i^*M = M|U$ for each sheaf M on X.

Exercise 169 Concrete
Let $f: S^1 \to S^1$ be the map $z \mapsto z^2$. Describe f_*C and f^*C, where C is the sheaf of continuous real-valued functions. What do the corresponding étale spaces $|f_*C|$ and $|f^*C|$ look like?

Exercise 170 Not Just Pushouts
Show that the functor f_* preserves all pullbacks.

Simplicial Sets 11

Simplicial sets are discrete, combinatorial models for topological spaces that simplify the construction of many objects in topology. It turns out that the homotopy theory of simplicial sets is equivalent to that of topological spaces. The importance of simplicial sets for topology is based on the fact that many objects in topology are constructed as realisations of simplicial sets. Some perspectives on this will be given at the end of this chapter.

11.1 Simplicial Objects and Morphisms

First, we will describe the *simplicial category* Δ. It codifies the combinatorial structure underlying the simplices.

Definition 11.1.1

For an integer $n \geqslant 0$, let $[n]$ be the small category that belongs to the ordered set

$$\{0 \leqslant 1 \leqslant \cdots \leqslant n\}$$

as described in Sect. 1.4. Let Δ be the full subcategory of the category of small categories comprising these categories $[n]$. The morphisms from $[m]$ to $[n]$ are all functors between the two categories.

▶ **Remark 11.1.2** The set of functors from $[m]$ to $[n]$ consists of the maps $f : [m] \to [n]$ that are monotonically increasing in the weak sense.

Since [0] has a unique element, there is a unique morphism $[n] \to [0]$. Thus, we have a *terminal object*, i.e., an empty product, in the category. Conversely, there are $n+1$ morphisms $[0] \longrightarrow [n]$. In general, there are

$$\binom{m+n+1}{m+1}$$

functors from $[m]$ to $[n]$. For each $i \in \{0, \ldots, n\}$, there are the *boundary maps*

$$d^i : [n-1] \longrightarrow [n], \quad x \longmapsto \begin{cases} x & x < i \\ x+1 & x \geqslant i, \end{cases}$$

which omit i, and the *degeneracy maps*

$$s^i : [n+1] \longrightarrow [n], \quad x \longmapsto \begin{cases} x & x \leqslant i \\ x-1 & x > i, \end{cases}$$

which hit i twice. The following *cosimplicial* relations between these morphisms are quickly checked:

$$\begin{aligned} d^j d^i &= d^i d^{j-1} && i < j \\ s^j s^i &= s^i s^{j+1} && i \leqslant j \\ s^j d^i &= \begin{cases} d^i s^{j-1} & i < j \\ \mathrm{id} & i = j, j+1 \\ d^{i-1} s^j & i > j+1. \end{cases} \end{aligned}$$

Theorem 11.1.3

Every morphism $f : [m] \to [n]$ in Δ has a unique presentation in the form

$$f = d^{i(r)} \ldots d^{i(1)} s^{j(1)} \ldots s^{j(s)},$$

where $i(1) < \cdots < i(r)$ and $j(1) < \cdots < j(s)$.

Proof. There is a splitting of f into the surjection onto the image and the inclusion of the image. Let

$$[n] \setminus \mathrm{Image}(f) = \{i(1) < \cdots < i(r)\}$$

and

$$\{j \in [m] \mid f(j) = f(j+1)\} = \{j(1) < \cdots < j(s)\}.$$

11.1 Simplicial Objects and Morphisms

Then

$$f: [m] \xrightarrow{\text{surj.}} [n-r] = [m-s] \xrightarrow{\text{inj.}} [n]$$

is the desired factorisation. □

Corollary 11.1.4
A functor on Δ is uniquely determined by the values on the d^i and s^i. On the other hand, it can also be prescribed by the images of these morphisms if the cosimplicial relations are fulfilled.

Proof. The uniqueness statement follows directly from the theorem. For the second statement, suppose that F preserves the cosimplicial relations. We must show that the expression

$$F(d^{i(1)}) \ldots F(d^{i(r)}) F(s^{j(1)}) \ldots F(s^{j(s)})$$

has the functoriality property. However, this follows from the observation that the unique factorisation of a composition of morphisms is obtained by applying the cosimplicial relations to the composition of the factorised morphisms. □

Definition 11.1.5

Let Δ^{op} be the category that arises from Δ by reversing the arrows (see Sect. 1.4). Let $\mathcal{C}$ be a category. Let $\Delta^{\text{op}}\text{-}\mathcal{C}$ denote the category of functors

$$F: \Delta^{\text{op}} \longrightarrow \mathcal{C}$$

together with the natural transformations as morphisms. The objects are also called *simplicial objects* in $\mathcal{C}$. The functor category

$$\Delta\text{-}\mathcal{C} = \text{Hom}(\Delta, \mathcal{C})$$

is the category of *cosimplicial objects* in $\mathcal{C}$.

Examples 11.1.6
If $\mathcal{C}$ is the category of small categories then the inclusion functor $\Delta \to \mathcal{C}$ is a cosimplicial category. If $\mathcal{C}$ is any category and C is an object in it then the constant functor with image C provides a simplicial object (and also a cosimplicial object). Such objects are also called *discrete*.

Now we define *simplicial sets* as simplicial objects in the category of sets. A simplicial set X is given by sets $X_n = X([n])$ and maps

$$d_i = X(d^i): X_n \longrightarrow X_{n-1}, \quad s_i = X(s^i): X_n \longrightarrow X_{n+1}$$

that fulfil the above relations in the reversed order. In detail, the simplicial relations are

$$d_i d_j = d_{j-1} d_i \quad i < j$$
$$s_i s_j = s_{j+1} s_i \quad i \leqslant j$$
$$d_i s_j = \begin{cases} s_{j-1} d_i & i < j \\ \text{id} & i = j, j+1 \\ s_j d_{i-1} & i > j+1. \end{cases}$$

However, we do not recommend memorising these relations to understand the nature of simplicial sets. Preferably, we work with all morphisms of the simplicial category Δ without using their presentation by generators and relations.

Definition 11.1.7

Let X be a simplicial set. Then X_n is the set of *n-simplices* in X. Their images under the above maps are the *boundaries* and the *degenerate simplices*. Morphisms between simplicial sets are called *simplicial maps*.

The boundary and degeneracy maps of the simplices can be remembered by the scheme

$$X_0 \rightrightarrows X_1 \substack{\longrightarrow \\ \longleftarrow \\ \longrightarrow} X_2 \substack{\longrightarrow \\ \longleftarrow \\ \longrightarrow \\ \longleftarrow} X_3 \cdots$$

Simplicial maps are maps between the sets of n-simplices for each n that are compatible with boundaries and degeneracies.

We now come to the definition of a sequence of particularly important simplicial sets. The adjoint of the morphism functor

$$\text{Mor}(?, ?): \Delta^{\text{op}} \times \Delta \longrightarrow \textbf{Sets}$$

is the embedding

$$\Delta \longrightarrow \Delta^{\text{op}}\text{-}\textbf{Sets}.$$

This results in a cosimplicial simplicial set.

11.1 Simplicial Objects and Morphisms

Definition 11.1.8

The image of $[n]$ is a simplicial set, denoted by Δ^n.

According to this definition, the m-simplices of Δ^n are given by

$$\mathrm{Mor}_\Delta([m],[n]) = \mathrm{Mor}_{\Delta^{\mathrm{op}}}([n],[m])$$

This is the set of (weakly) increasing maps

$$\{0 \leqslant \cdots \leqslant m\} \longrightarrow \{0 \leqslant \cdots \leqslant n\}.$$

The 0-simplices of Δ^n are thus the $n+1$ objects $0,\ldots,n$ of $[n]$; the 1-simplices of Δ^n are the morphisms of $[n]$, i.e., the pairs (i,j) of integers with $0 \leqslant i \leqslant j \leqslant n$; the 2-simplices of Δ^n are the pairs of composable morphisms of $[n]$ and so on.

Example 11.1.9
If $n=1$, the number of m-simplices of Δ^1 is $m+2$. The two 0-simplices are 0 and 1. The m-simplices for $m \geqslant 1$ are sequences

$$0 \to 0 \to 0 \to \cdots \to 0 \to 1 \to \cdots \to 1 \to 1$$

of m composable morphisms in $[1]$. The boundary and degeneracy maps are given by composition (deletion of objects) and insertion of identities (duplication of objects).

Much more significant than the 'inner life' given by the simplices and structure maps of Δ^n is its following property:

Theorem 11.1.10
The morphisms from Δ^n to a simplicial set X are in one-to-one correspondence with the n-simplices X_n of X by means of the bijection

$$\mathrm{Mor}(\Delta^n, X) \cong X_n, \quad \varphi \longmapsto \varphi_{[n]}(\mathrm{id}_{[n]}).$$

Proof. The proof follows from very general considerations that bear the name *Yoneda Lemma*: if $F \colon \mathcal{C} \to \mathbf{Sets}$ is a functor and C is an object of $\mathcal{C}$ then the natural map

$$\mathrm{Mor}(\mathrm{Mor}(C,?), F) \longrightarrow F(C), \quad \Phi \longmapsto \Phi_C(\mathrm{id}_C)$$

is a bijection. This is quickly verified for each $f: C \to D$ using the commutative diagram

$$\begin{array}{ccc} \mathrm{id}_C & \mathrm{Mor}(C,C) & \xrightarrow{\Phi_C} F(C) \\ \downarrow & {\scriptstyle \mathrm{Mor}(C,f)}\downarrow & \downarrow{\scriptstyle F(f)} \\ f & \mathrm{Mor}(C,D) & \xrightarrow{\Phi_D} F(D). \end{array}$$

$\square$

Definition 11.1.11

The *product* of simplicial sets X and Y is defined by

$$(X \times Y)_n = X_n \times Y_n$$

and the product of the boundary and degeneracy maps. The *simplicial mapping set* is

$$\mathrm{Hom}(X,Y)_n = \mathrm{Hom}(\Delta^n \times X, Y)$$

together with the simplicial structure induced by the cosimplicial simplicial set $n \mapsto \Delta^n$.

Theorem 11.1.12

There is a natural bijection

$$\mathrm{Mor}(X \times Y, Z) \to \mathrm{Mor}(X, \mathrm{Hom}(Y,Z)), \ f \mapsto f^{\#}$$

for all simplicial sets X, Y, and Z.

Proof. Let $f: X \times Y \to Z$ and a simplex x of X_n be given. Then the adjoint map $f^{\#}(x)$ is the composite

$$\Delta^n \times Y \xrightarrow{x \times \mathrm{id}} X \times Y \xrightarrow{f} Z \ .$$

Here, the previous theorem was used. This map is simplicial and provides the desired bijection. $\square$

Exercises

Exercise 171 General Nonsense
For every simplicial map $\varphi \colon \Delta^n \to \Delta^m$ there is a unique $f \colon [n] \to [m]$ in Δ such that $\varphi = \Delta(?, f)$ holds.

Exercise 172 Name Justified
Show that the product of simplicial sets possesses the universal property.

Exercise 173 Exponential Laws
Show that the above-stated adjunction provides an isomorphism

$$\mathrm{Hom}(X \times Y, Z) \cong \mathrm{Hom}(X, \mathrm{Hom}(Y, Z))$$

of simplicial sets.

11.2 Singular Simplices and Realisations

In this section, the singular complexes of topological spaces and the realisation of simplicial sets are introduced. They form an adjoint pair. This is the starting point of all comparisons between the categories of topological spaces and simplicial sets. The basis is the cosimplicial space formed from the standard simplices.

Definition 11.2.1

Let $n \geqslant 0$ be an integer. The subspace

$$\Delta_{\mathrm{top}}^n = \{(t_0, \ldots, t_n) \in \mathbb{R}^{n+1} \mid t_i \geqslant 0 \text{ for all } i = 0, \ldots, n \text{ and } \sum_{i=0}^n t_i = 1\}$$

is called the *n-dimensional standard simplex*.

The space Δ_{top}^n is thus a compact convex subset of $\mathbb{R}^{n+1}$. The canonical basis vectors $e_0, \ldots, e_n$ of $\mathbb{R}^{n+1}$ are its $n+1$ vertices, and $\frac{1}{n+1} \sum_{i=0}^n e_i$ is its centroid (see Fig. 11.1).

We get continuous maps $\Delta_{\mathrm{top}}^n \to \mathbb{R}^k$ by restricting linear maps $\mathbb{R}^{n+1} \to \mathbb{R}^k$. To specify these, it suffices to specify the images of the basis $\{e_0, \ldots, e_n\}$. In particular,

Fig. 11.1 The space Δ_{top}^n for $n = 0, 1, 2, 3$

we have continuous maps

$$\Delta_{\text{top}}^{n-1} \longrightarrow \Delta_{\text{top}}^n, e_x \mapsto e_{d^i(x)}$$

and

$$\Delta_{\text{top}}^{n+1} \longrightarrow \Delta_{\text{top}}^n, e_x \mapsto e_{s^i(x)}$$

from the boundary and degeneracy maps of the category Δ.

▶ **Remark 11.2.2** The assignment $[n] \mapsto \Delta_{\text{top}}^n$ together with the maps just mentioned provides a covariant functor $\Delta \to \textbf{Top}$, thus a cosimplicial topological space.

This allows the following definition.

Definition 11.2.3

Let X be any topological space. Then the assignment

$$[n] \mapsto \text{Sing}_n(X) = \text{Mor}(\Delta_{\text{top}}^n, X)$$

is a simplicial set $\text{Sing}(X)$. The elements of $\text{Sing}_n(X)$ are called *singular n-simplices* in X.

The construction itself is functorial, thus giving us a functor

$$\text{Sing} \colon \textbf{Top} \longrightarrow \Delta^{\text{op}}\textbf{-Sets}.$$

Topological spaces thus provide (important!) examples for simplicial sets.

Next, we want to assign a topological space to a simplicial set X.

Definition 11.2.4

The *realisation* $|X|$ of a simplicial set X is the quotient space obtained from

$$\coprod_{n \geq 0} X_n \times \Delta_{\text{top}}^n$$

by the relation

$$(X_f(x), t) \sim (x, \Delta_{\text{top}}^f(t))$$

11.2 Singular Simplices and Realisations

for all morphisms $f: [m] \to [n]$ in Δ. Here, the set X_n carries the discrete topology. (If X had been a simplicial topological space, we would use the topology of X_n for its realisation in the quotient formation.)

Example 11.2.5
The realisation of the simplicial set Δ^n is homeomorphic to Δ^n_{top}. To see this, note that the realisation of a simplicial set is homeomorphic to the space that arises from the sum of all nondegenerate n-simplices $\{x\} \times \Delta^n_{\text{top}}$ by identification of the boundaries. In Δ^n, only an n-simplex and its boundaries are nondegenerate. This implies the assertion.

A simplicial map $\varphi: X \to Y$ provides a continuous map $|\varphi|: |X| \longrightarrow |Y|$. Thus, the realisation is a functor

$$|?|: \Delta^{\text{op}}\text{-}\mathbf{Sets} \longrightarrow \mathbf{Top}.$$

Every simplicial set thus describes a topological space. An important aspect of the relationships between the realisation and the singular simplicial set is provided by the following result.

Theorem 11.2.6
The functors $|?|$ and $\text{Sing}(?)$ are adjoint: For every simplicial set X and every topological space Y there is a natural bijection

$$\text{Mor}(|X|, Y) \cong \text{Mor}(X, \text{Sing}(Y)).$$

Proof. Each $f: |X| \to Y$ defines, for each simplex x of X_n, a singular simplex of Y by the composite

$$\Delta^n_{\text{top}} \cong \{x\} \times \Delta^n_{\text{top}} \longrightarrow |X| \xrightarrow{f} Y .$$

The identifications in the quotient space $|X|$ ensure this assignment is simplicial. Conversely, given a simplicial map φ from X into the singular simplicial set of Y, a continuous map on the realisation is defined by

$$[(x, t)] \mapsto \varphi(x)(t).$$

These assignments are inverses of each other. □

The realisations of simplicial sets are always very special topological spaces. For example:

▶ **Remark 11.2.7** Realisations of simplicial sets are always locally-compactly generated.

Because of the universal property of the quotient space, a map from a realisation is continuous if and only if it is continuous when restricted to the individual simplices. Since they are locally compact, realisations are locally-compactly generated. We can also show that realisations are always cell complexes (see Exercise 175), so in particular, they have the Hausdorff property. Realisations of products are the products of the realisations if we take an appropriate product to stay in the category of locally-compactly generated spaces:

Theorem 11.2.8
The canonical map

$$(|\mathrm{pr}_X|, |\mathrm{pr}_Y|) \colon |X \times Y| \longrightarrow |X| \times_k |Y|$$

is a homeomorphism for all simplicial sets X and Y.

Proof. The proof is divided into several steps.

(Step 1) First, consider the case $X = \Delta^m$ and $Y = \Delta^n$. It will turn out that the general case can be reduced to this one. We proceed in the following steps. Let us first try to understand the non-degenerate simplices of $\Delta^m \times \Delta^n$. A k-simplex in this product is a weakly increasing map $[k] \to [m] \times [n]$, where $(a, b) \leqslant (a', b')$ in $[m] \times [n]$ if $a \leqslant a'$ and $b \leqslant b'$ holds. It is non-degenerate if the map is injective or strictly monotone. There are obviously only finitely many and none as soon as $k > m + n$. Therefore, the space $|\Delta^m \times \Delta^n|$ is compact, and it suffices to prove the bijectivity of the canonical map. A non-degenerate k-simplex in $\Delta^m \times \Delta^n$ describes a chain of points in the grid between $(0, 0)$ and (m, n). A chain shall be called *maximal* if $k = m + n$ holds. Every chain can be extended to a maximal chain by adding suitable points as in Fig. 11.2. Therefore, for $k < n + m$, every k-simplex is the boundary of a non-degenerate $(m+n)$-simplex. Thus, in the realisation, we can focus on such. In addition, there are exactly two such extensions of a chain to two $(m+n)$-simplices that intersect in the present side. Each point of a maximal chain is numbered with the numbers $0, 1, \ldots, m + n$. It is completely determined by the numbers on the ends of the m horizontal segments. For example, the triple $(1, 4, 6)$ describes the maximal chain of the diagram. Conversely each m-element subset of $\{1, 2, \ldots, m + n\}$ describes a maximal chain. Thus, there are $\binom{m+n}{m}$ maximal chains.

For example, for $m = n = 1$, we have the two 2-simplices $(1) = \{(0, 0), (0, 1), (1, 1)\}$ and $(2) = \{(0, 0), (1, 0), (1, 1)\}$. These intersect in the 1-simplex corresponding to the chain $\{(0, 0), (1, 1)\}$.

11.2 Singular Simplices and Realisations

Fig. 11.2 A chain and a maximal extension

In this notation, the restriction of the canonical map to an $(n+m)$-simplex for a maximal chain $(a_1, a_2, \ldots, a_m)$ has the following form. Let $(b_1, b_2, \ldots, b_n)$ be the complementary maximal chain, which is obtained by reflection on the diagonal. Then an element x of $|\Delta^{m+n}|$ is mapped to $(y, z) \in |\Delta^m| \times |\Delta^n|$, where

$$y_j = \sum_{k=a_j}^{a_{j+1}-1} x_k \quad \text{and} \quad z_j = \sum_{k=b_j}^{b_{j+1}-1} x_k$$

for $k \geq 1$ and $a_{m+1} = m+n+1 = b_{n+1}$. This map is injective in any case. It remains to show that every point (y, z) of $|\Delta^m| \times |\Delta^n|$ is in the image of a $(n+m)$-simplex, and that any two such $(n+m)$-simplices intersect in a side that contains the point: set $u_j = y_j + \ldots + y_m$ and $v_j = z_j + \ldots + z_n$ for $j \geq 1$. Arrange these numbers in descending order $w_1 \geq w_2 \geq \cdots \geq w_{m+n}$. There are several possibilities if equal numbers occur in this sequence. Each u_j is then a w_{a_j}. The numbers $(a_1, a_2, \ldots, a_m)$ define a maximum chain and thus a $n+m$-simplex. The coordinates $x_j = w_j - w_{j+1}$ in this simplex map to (y, z) and the possibilities of the orders correspond to the assignments of side points to the $n+m$-simplices.

(Step 2) Let $\mathcal{C}$ be the class of all simplicial sets X for which the canonical map

$$|X \times \Delta^n| \longrightarrow |X| \times_k |\Delta^n|$$

is a homeomorphism. It was shown in the first step that all standard simplices Δ^m are in $\mathcal{C}$. The class $\mathcal{C}$ is also closed with respect to pushouts, i.e., if

$$\begin{array}{ccc} X_0 & \longrightarrow & X_1 \\ \downarrow & & \downarrow \\ X_2 & \longrightarrow & X \end{array}$$

is a pushout of simplicial sets with $X_j \in \mathcal{C}$ for $j = 0, 1, 2$ then $X \in \mathcal{C}$. To check this, we only need to note that both the realisation functor and the product with Δ^n in Δ^{op}-**Sets** and with $|\Delta^n|$ in **Top**, preserve pushouts (due to the exponential laws). For similar reasons, the class $\mathcal{C}$ is also closed with respect to all sums; if we have $X_j \in \mathcal{C}$, then we have also have $\coprod_j X_j \in \mathcal{C}$:

$$\left| \coprod_j X_j \times \Delta^n \right| \cong \coprod_j |X_j \times \Delta^n| \cong \coprod_j |X_j| \times_k |\Delta^n| \cong \left| \coprod_j X_j \right| \times_k |\Delta^n|.$$

We claim that, using this, all simplicial sets X must already be in $\mathcal{C}$: let $\mathcal{S}(X)$ be the category of all simplicial maps $\Delta^n \to X$. These correspond to the n-simplices of X according to Theorem 11.1.10. Morphisms in $\mathcal{S}(X)$ are commutative triangles

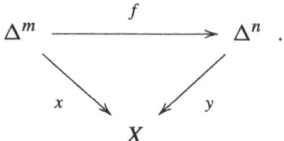

Then every X sits in a pushout diagram of the form

$$\begin{array}{ccc} \coprod_{x:\,\Delta^n\to X} \Delta^n + \coprod_{f:\,x\to y} \Delta^m & \xrightarrow{(\mathrm{id},i)} & \coprod_{x:\,\Delta^n\to X} \Delta^n \\ {\scriptstyle (\mathrm{id},j)}\downarrow & & \downarrow \\ \coprod_{x:\,\Delta^n\to X} \Delta^n & \longrightarrow & X. \end{array}$$

The maps to X are given on each summand by the respective objects of $\mathcal{S}(X)$ itself. The map i is the horizontal arrow f of the underlying commutative triangle in the summand given by y. The map j is the identity map in the summands x. The reader is left to check the universal property of the pushout.

(Step 3) The second step showed that the standard simplices belong to the class of simplicial sets Y for which

$$|X \times Y| \longrightarrow |X| \times_k |Y|$$

is a homeomorphism for all simplicial sets X. Now, we can argue precisely as in the second step that all simplicial sets Y are in this class. □

Supplements

More About Realisation and Singular Simplices We would like to turn the adjunction of the realisation functor and the functor Sing into an equivalence between suitable homotopy theories. The realisations, however, are very particular spaces, and we cannot expect that every topological space is homotopy equivalent to a realisation. Conversely, not all simplicial sets are of the form $\mathrm{Sing}(X)$ for a topological space X. To describe the relationship between a space X and the realisation $|\mathrm{Sing}(X)|$ more precisely, the following term is introduced: let $f: X \to Y$ be a continuous map between topological spaces. Then f is called *weak homotopy equivalence*, if the induced map on the path components and homotopy groups

$$f_*: \pi_n(X, x) \longrightarrow \pi_n(Y, f(x))$$

11.2 Singular Simplices and Realisations

is an isomorphism for all $x \in X$ and all $n \geqslant 0$. For example, every homotopy equivalence is a weak equivalence. The converse is not true: the sinusoidal space from Sect. 3.1 is not homotopy equivalent to S^0 because it is connected. However, if we choose a point in both path components, the map from S^0 is a weak homotopy equivalence. Weak homotopy equivalences are always homotopy equivalences if the involved spaces can be built from cells; this is a theorem by Whitehead (see [Hat02]). The following theorem by Quillen is more astonishing (for a proof see [GJ99]): for every topological space X, the adjoint

$$|\mathrm{Sing}(X)| \longrightarrow X$$

to the identity of $\mathrm{Sing}(X)$ is a weak homotopy equivalence. In particular, no information is lost, up to weak homotopy equivalence, when transitioning from topological spaces to simplicial sets and back to topological spaces.

Kan Fibrations The concept of a fibration can be transferred to simplicial sets. For $0 \leqslant r \leqslant n$, let the r-th *horn* Λ_r^n be the simplicial subset of Δ^n whose m-simplices are given by order-preserving maps $f\colon [m] \to [n]$ for which r is not in the image. (Geometrically, we obtain $|\Lambda_r^n|$ from $|\Delta^n|$ by removing the interior of $|\Delta^n|$ and the interior of the side opposite to the vertex r. For $n = 2$, this then looks like a horn Λ.) A map $p\colon X \to Y$ between simplicial sets is called a *fibration* (or more precisely a *Kan fibration* or *Kan–Quillen fibration*) if for each r and each commutative diagram

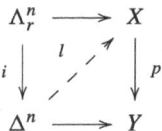

of simplicial sets, a lifting exists that makes the two triangles commutative. A simplicial set X is called *fibrant* if the map $X \to \Delta^0 = \star$ is a fibration. This definition surprises in two places: it does not seem to be enough for fibrations in the combinatorial world to consider the inclusion of one side of a cube or the inclusion of a single horn. Moreover, the map into the point object is not necessarily a fibration. For example, the simplicial sets Δ^n for $n \geqslant 1$ are not fibrant. Figure 11.3 depicts a morphism f that cannot be extended to Δ^2.

An example of a fibrant object is $\mathrm{Sing}(X)$ for every topological space X. This follows from the following observation: a map $f\colon X \to Y$ between topological spaces is a Serre fibration if and only if $\mathrm{Sing}(f)\colon \mathrm{Sing}(X) \to \mathrm{Sing}(Y)$ is a Kan fibration. To see this, we only need to note that the inclusion $|\Lambda_r^n| \subseteq \Delta^n$ is

Fig. 11.3 Δ^2 is not fibrant

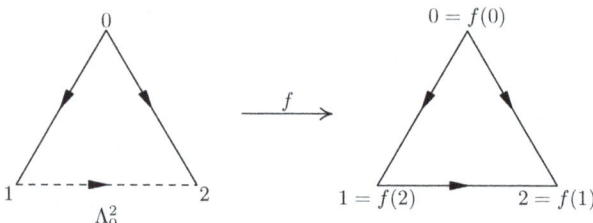

homeomorphic to $\{0\} \times I^n \subseteq I^n$ and then exploit the adjunction of the diagrams

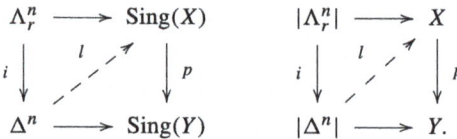

Simplicial Mapping Spaces Let Y and Z be topological spaces. Then we already know the space $\mathrm{Hom}(Y, Z)$ of continuous maps from Y to Z with the compact-open topology. A singular n-simplex in $\mathrm{Hom}(Y, Z)$ is a continuous map

$$\Delta_{\mathrm{top}}^n \to \mathrm{Hom}(Y, Z).$$

If Y is locally compact, this is equivalent to a map

$$\Delta_{\mathrm{top}}^n \times Y \to Z.$$

This suggests defining a *simplicial mapping space* by

$$[n] \mapsto \mathrm{Mor}(\Delta_{\mathrm{top}}^n \times Y, Z)$$

Since the Δ_{top}^n form a cosimplicial space, the result is a simplicial set. We can now either continue with this or realise it to be able to work again in the category of topological spaces. For locally compact sources, this yields the old mapping space with the compact-open topology. In the general case, we obtain something closer to the intuition: the 0-simplices are the continuous maps, the 1-simplices are the homotopies between such and so on.

Exercises

Exercise 174 Spherical
Let $\partial \Delta^n \subseteq \Delta^n$ be the simplicial subset whose m-simplices consist of the non-surjective, weakly monotone maps from $[m]$ to $[n]$. Show that the realisation $|\partial \Delta^n|$ is homeomorphic to the boundary of the standard simplex $\Delta_{\mathrm{top}}^n \subseteq \mathbb{R}^{n+1}$.

Exercise 175 Skeletons and Cells
For a simplicial set X, let the *n-skeleton*
$$\mathrm{sk}_n(X) \subseteq X$$
be the simplicial subset that contains all simplices of degree $\leq n$ and their degeneracies. Let $N(X_n) \subseteq X_n$ denote the set of non-degenerate simplices. Show that there is a pushout diagram of the form

$$\begin{array}{ccc} \coprod_{x \in N(X_n)} \partial \Delta^n & \longrightarrow & \mathrm{sk}_{n-1}(X) \\ \subseteq \downarrow & & \subseteq \downarrow \\ \coprod_{x \in N(X_n)} \Delta^n & \longrightarrow & \mathrm{sk}_n(X), \end{array}$$

where $\partial \Delta^n$ is defined as in the preceding exercise. Conclude from this that $|\mathrm{sk}_n(X)|$ arises from $|\mathrm{sk}_{n-1}(X)|$ by attaching n-cells.

11.3 Outlook

We conclude this chapter (and thus this book) with some outlooks on results that can be accomplished with the presented techniques. We will (in the style of the previous supplements) again not prove all statements in detail. Rather, we want to indicate that, with the end of this book, topology is far from being over. On the contrary, we hope the reading of this book is only the beginning of a closer engagement with topology.

Singular Homology and Cohomology If X is a set, then we can consider the free abelian group

$$\mathbb{Z}[X] = \bigoplus_X \mathbb{Z}$$

generated by this set. An element is a linear combination

$$\sum_{x \in X} \lambda_x \cdot x$$

with $\lambda_x \in \mathbb{Z}$ and $\lambda_x = 0$ for almost all $x \in X$. This provides a functor

$$\mathbb{Z}[\cdot]: \mathbf{Sets} \longrightarrow \mathbf{AbGrp}$$

into the category of abelian groups. It is left adjoint to the forgetful functor, i.e., we have

$$\mathrm{Abb}(X, A) \cong \mathrm{Hom}_{\mathbb{Z}}(\mathbb{Z}[X], A)$$

for every abelian group A. A homomorphism $\mathbb{Z}[X] \to A$ is uniquely determined by its values on the basis, that is, on the points of X. Conversely, every assignment of elements of X also provides a unique homomorphism. This functor is the gateway to the transition from the nonlinear categories of sets, simplicial sets and topological spaces into the linear (more precisely, additive) category of (simplicial) abelian groups. By composing with this functor, we obtain a functor

$$\mathbb{Z}[\cdot] \colon \Delta^{\mathrm{op}}\text{-}\mathbf{Sets} \longrightarrow \Delta^{\mathrm{op}}\text{-}\mathbf{AbGrp}$$

into the category of simplicial abelian groups. By composing with the singular complex a functor, we get a functor

$$\mathbb{Z}[\mathrm{Sing}(\cdot)] \colon \mathbf{Top} \longrightarrow \Delta^{\mathrm{op}}\text{-}\mathbf{AbGrp},$$

called *linearisation*. As explained in Sect. 11.2, no homotopy-theoretically relevant information is lost when transitioning from a topological space X to its simplicial complex. Only the linearisation is, as it turns out, a crude simplification. Simplicial abelian groups do not seem easy to handle objects at first glance. We can approach them in two equivalent ways. On the one hand, a simplicial abelian group becomes a pointed simplicial set by forgetting the group structure. We can realise this and thus obtain from X a new topological space $|\mathbb{Z}[\mathrm{Sing}(X)]|$. Its homotopy groups are then invariants of X. On the other hand, simplicial abelian groups also have their own homotopy theory, making them equivalent to the category of (non-negatively graded) chain complexes of abelian groups. Thus, again, no homotopically relevant information is lost. The equivalence in one direction is given as follows: given A is a simplicial abelian group, a complex of abelian groups can be formed

$$0 \longleftarrow \mathbb{Z}[A_0] \xleftarrow{\partial_1} \mathbb{Z}[A_1] \xleftarrow{\partial_2} \mathbb{Z}[A_2] \xleftarrow{\partial_3} \cdots$$

by setting

$$\partial_n = \sum_{i=0}^{n} (-1)^i \mathbb{Z}[d^i]$$

It follows from the simplicial identities of Sect. 11.1 that $\partial\partial = 0$ holds. So this sequence is a chain complex of abelian groups. With this chain complex, we can carry out various algebraic constructions. For example, we can define the homology as follows:

$$H_n(X; \mathbb{Z}) = \frac{\mathrm{Kern}(\partial_n \colon \mathbb{Z}[\mathrm{Sing}(X)_n] \to \mathbb{Z}[\mathrm{Sing}_{n-1}(X)])}{\mathrm{Im}(\partial_{n+1} \colon \mathbb{Z}[\mathrm{Sing}(X)_{n+1}] \to \mathbb{Z}[\mathrm{Sing}_n(X)])}.$$

We call this the *singular homology groups* of a space X. This algebraic construction calculates the above-mentioned homotopy groups in the sense that there is an isomorphism

$$H_n(X; \mathbb{Z}) \cong \pi_n(|\mathbb{Z}[\mathrm{Sing}(X)]|, 0)$$

of abelian groups. For proof of this result, we refer to [GJ99]. The complex can also be tensored with other abelian groups (then we speak of *coefficients*), or it can be mapped into an abelian group (then we speak of *cohomology*). However, we should not overlook that further information can be lost in the process. Ideally, we should, therefore, continue working with the complex itself so that the linearisation is the only process that is homotopically relevant. Homology theory is the theory of this linearisation.

Classifying Spaces We now explain how we can assign a space to each small category (see Sect. 1.4), a continuous map to each functor, and a homotopy to each natural transformation. Small categories can thus be used to describe (important!) spaces. Let **Cat** be the category of small categories. The adjoint to the morphism functor

$$\mathrm{Mor}(?, ??) \colon \Delta^{\mathrm{op}} \times \mathbf{Cat} \longrightarrow \mathbf{Sets}$$

is a functor

$$\mathrm{B} \colon \mathbf{Cat} \longrightarrow \Delta^{\mathrm{op}}\text{-}\mathbf{Sets}$$

that assigns to a small category $\mathcal{C}$ its *nerve* $\mathrm{B}\mathcal{C}$. Another name for this is *classifying space*. For example, we have $\mathrm{B}[n] = \Delta^n$, which should be obvious with the given definition. In the general case, the n-simplices of $\mathrm{B}\mathcal{C}$ are just the functors from $[n]$ to $\mathcal{C}$. In particular, the 0-simplices of $\mathrm{B}\mathcal{C}$ are the objects of $\mathcal{C}$, the 1-simplices of $\mathrm{B}\mathcal{C}$ are the morphisms of $\mathcal{C}$, and the 2-simplices of $\mathrm{B}\mathcal{C}$ are the pairs of composable morphisms of $\mathcal{C}$. If the set of objects is denoted with U and the set of morphisms with R, then we have

$$\mathrm{B}\mathcal{C}_n = R \times_U R \times_U \cdots \times_U R,$$

with n copies of R and we can understand the structure maps.

The functor B commutes with products:

$$\mathrm{B}(\mathcal{C} \times \mathcal{D}) \cong \mathrm{B}\mathcal{C} \times \mathrm{B}\mathcal{D}.$$

This follows directly from the universal properties. It also results from the fact that B has a left adjoint. Every functor $\mathcal{C} \to \mathcal{D}$ induces a simplicial map $\mathrm{B}\mathcal{C} \to \mathrm{B}\mathcal{D}$ due to the functoriality of B. Since a natural transformation between two functors $\mathcal{C} \to \mathcal{D}$ is the same as a functor $\mathcal{C} \times [1] \to \mathcal{D}$, a natural transformation provides a homotopy

$$\mathrm{B}\mathcal{C} \times \Delta^1 = \mathrm{B}(\mathcal{C} \times [1]) \longrightarrow \mathrm{B}\mathcal{D}.$$

It follows that equivalent categories have homotopy equivalent classifying spaces. A category with an initial object (i.e., an empty sum) or terminal object (i.e., an empty product) has a contractible classifying space.

If $\mathcal{G}$ is a groupoid (or even a group), then the set $\pi_0(|B\mathcal{G}|)$ of path components is the set of isoclasses of objects of $\mathcal{G}$ (see [GJ99]). If X is an object of $\mathcal{G}$, there is an isomorphism

$$\pi_1(|B\mathcal{G}|, X) \cong \operatorname{Aut}(X)$$

of groups. The inclusion of $\operatorname{Aut}(X)$ into $\mathcal{G}$ induces an equivalence of categories onto the component of X in $B\mathcal{G}$. Summation thus yields a homotopy equivalence

$$|B\mathcal{G}| \simeq \coprod_{[X]} |B\operatorname{Aut}(X)|,$$

where the sum runs over a representative system of the isomorphism classes of objects X of $\mathcal{G}$.

Simplicial Resolutions Let $f: U \to X$ be a continuous map between topological spaces. Consider the (topological) category $\mathcal{R}_f$ with object space U and a morphism $u_1 \to u_2$ whenever $f(u_1) = f(u_2)$. It follows for the 1-simplices of the classifying space $B\mathcal{R}_f$ that

$$(B\mathcal{R}_f)_1 = U \times_X U.$$

This is the equivalence relation associated with f. Conversely, if an equivalence relation on U is given, we have the projection from U onto the set of equivalence classes. For example, if $(U_j \mid j \in J)$ is a cover of X then

$$U = \coprod_j U_j$$

maps to X in an obvious way, and we have

$$U \times_X U = \coprod_{i,j} (U_i \cap U_j).$$

We should interpret $B\mathcal{R}_f$ as *a resolution of X using U* because of the morphism of simplicial spaces

$$\begin{array}{c} U \rightrightarrows U \times_X U \substack{\rightarrow \\ \rightarrow \\ \rightarrow} \cdots \\ \downarrow \\ X \end{array}.$$

11.3 Outlook

Here, the space X is considered as a discrete simplicial space. (Its realisation is again X.) The two extreme examples are worth looking at: the constant map $k\colon U \to \star$ corresponds to the equivalence relation where all points are equivalent. We have

$$(B\mathcal{R}_k)_n = U^{n+1},$$

and it is easy to see that (the realisation of) $B\mathcal{R}_k$ is contractible. Think of it as a thickening of the point $\star$ by means of U. The other example is the identity $\mathrm{id}\colon U \to U$. It corresponds to the equivalence relation where each point is only equivalent to itself. We have

$$(B\mathcal{R}_{\mathrm{id}})_n \cong U,$$

for all n, and $B\mathcal{R}_{\mathrm{id}}$ is the constant simplicial space with value U. Its realisation is equivalent to U. As a last example, consider an action of a group G on a space U. It provides a map $p\colon U \to U/G = X$ and thus a resolution. The action of G on itself by left translation results in the resolution of $\star = G/G$ through a contractible space EG, on which G acts freely.

Nerves of Coverings Let X be a topological space, and let $(U_j \mid j \in J)$ be a cover of X. For a subset $I \subseteq J$, let

$$U_I = \bigcap_{i \in I} U_i,$$

so that in particular $U_{\{i\}} = U_i$. Let $\mathcal{U}$ be the category of the non-empty U_I with finite I and their inclusions. We have

$$B\mathcal{U}_n \cong \{\, (i_0, \ldots, i_n) \mid U_{\{i_0, \ldots, i_n\}} \neq \emptyset \,\}$$

and

$$B\mathcal{U}_f(i_0, \ldots, i_n) = (i_{f(0)}, \ldots, i_{f(n)}).$$

If $(U_j \mid j \in J)$ is a particularly nice cover, then

$$|B\mathcal{U}| \simeq X.$$

For a proof, see [Seg68, Prop. (4.1)]. It is also worth looking at [BT82], not just for this.

Bibliography

[BT82] Bott, R., & Tu, L. W. (1982). *Differential forms in algebraic topology. Graduate texts in mathematics* (Vol. 82). Springer-Verlag.

[Brö03] Bröcker, T. (2003). *Lineare Algebra und Analytische Geometrie. Ein Lehrbuch für Physiker und Mathematiker.* Birkäuser.

[BtD95] Bröcker, T., & tom Dieck, T. (1995). *Representations of compact lie groups. Graduate texts in mathematics* (Vol. 98). Springer-Verlag.

[BZ03] Burde, G., & Zieschang, H. (2003). *Knots. de Gruyter studies in mathematics* (2nd ed., Vol. 5). Walter de Gruyter & Co.

[Dol72] Dold, A. (1972). *Lectures on algebraic topology. Grundlehren der Mathematischen Wissenschaften* (Vol. 2, Auflage 1980). Springer-Verlag, neu 2004 in der Reihe Classics in Mathematics.

[Dre95] Dress, A. W. M. (1995). One more shortcut to Galois theory. *Advances in Mathematics, 110*, 129–140.

[Eng68] Engelking, R. (1968). *Outline of general topology.* North Holland.

[Ern74] Erné, M. (1974). Struktur- und Anzahlformeln für Topologien auf endlichen Mengen. *Manuscripta Mathematica, 11*, 221–259.

[GJ99] Goerss, P., & Jardine, J. F. (1999). *Simplicial homotopy theory. Progress in mathematics* (Vol. 174). Birkhäuser Verlag.

[GL89] Gordon, C. McA., & Luecke, J. (1989). Knots are determined by their complements. *Journal of the American Mathematical Society, 2*, 371–415.

[Gro12] Groves, J. (2012). An elementary counterexample in the compact-open topology. *The American Mathematical Monthly, 119*, 693–694.

[Hat02] Hatcher, A. (2002). *Algebraic topology.* Cambridge University Press.

[Hus94] Husemoller, D. (1994). *Fibre bundles* (3rd ed., Vol. 20). *Graduate texts in mathematics.* Springer-Verlag.

[Jän05] Jänich, K. (2005). *Topologie* (Achte Auflage). Springer-Verlag.

[Kir78] Kirby, R. C. (1978). A calculus for framed links in S^3. *Inventiones Mathematicae, 45*, 35–56.

[Kir89] Kirby, R. C. (1989). *The topology of 4-manifolds. Lecture Notes in Mathematics* (Vol. 1374). Springer-Verlag.

[Kre10] Kreck, M. (2010). *Differential algebraic topology. From stratifolds to exotic spheres. Graduate studies in mathematics* (Vol. 110). American Mathematical Society.

[LS23] Laures, G., & Szymik, M. (2023). *Grundkurs Topologie* (3rd ed.). Springer Spektrum Berlin.

[Lüc05] Lück, W. (2005). *Algebraische Topologie. Homologie und Mannigfaltigkeiten. Vieweg Studium: Aufbaukurs Mathematik.* Vieweg.

[Mac98] Mac Lane, S. (1998). *Categories for the working mathematician. Graduate texts in mathematics* (2nd ed., Vol. 5). Springer-Verlag.

[MM94] Mac Lane, S., & Moerdijk, I. (1994). *Sheaves in geometry and logic. A first introduction to topos theory.* Springer-Verlag.
[McC69] McCord, M. C. (1969). Classifying spaces and infinite symmetric products. *Transactions of the American Mathematical Society*, *146*, 273–298.
[Mil56] Milnor, J. W. (1956). On manifolds homeomorphic to the 7-sphere. *Annals of Mathematics*, *64*, 399–405.
[Mil65] Milnor, J. W. (1965). *Lectures on the h-cobordism theorem. Notes by L. Siebenmann and J. Sondow.* Princeton University Press.
[Oss92] Ossa, E. (1992). *Topologie. Vieweg Studium: Aufbaukurs Mathematik* (Vol. 42). Friedr. Vieweg & Sohn.
[Sch64] Schubert, H. (1964). *Topologie* (Vol. 1). Teubner.
[Seg68] Segal, G. (1968). Classifying spaces and spectral sequences. *Publications Mathématiques de l'Institut des Hautes Scientifiques*, No. 34, 105–112.
[ST34] Seifert, H., & Threlfall, W. (1934). *Lehrbuch der Topologie.* Teubner.
[Ser03] Serre, J.-P. (2003). On a theorem of Jordan. *Bulletin of the American Mathematical Society*, *40*, 429–440.
[SS95] Steen, L. A., & Seebach, J. A. Jr. (1995). *Counterexamples in topology.* Dover Publications, Inc.
[SZ94] Stöcker, R., & Zieschang, H. (1994). *Algebraische Topologie. Eine Einführung. Mathematische Leitfäden* (Zweite Auflage). B.G. Teubner.
[Sto66] Stong, R. E. (1966). Finite topological spaces. *Transactions of the American Mathematical Society*, *123*, 325–340.
[Szy14] Szymik, M. (2015). Homotopies and the universal fixed point property. *Order*, *32*, 301–311.
[tD87] tom Dieck, T. (1987). *Transformation groups, de Gruyter studies in mathematics* (Vol. 8). Walter de Gruyter & Co.
[tD91] tom Dieck, T. (1991). *Topologie. Erste Auflage. de Gruyter Lehrbuch.* Walter de Gruyter & Co.
[vQ79] von Querenburg, B. (1979). *Mengentheoretische Topologie.* Springer-Verlag.
[Whi37] Whitney, H. (1937). On regular closed curves in the plane. *Compositio Mathematica*, *4*, 276–284.

Index

A
Action, 86
   conjugation, 87
   effective, 187
   faithful, 187
   free, 88
   proper, 96
   transitive, 91
   trivial, 87
Adjoint
   of a functor, 213
   of a map, 75
Atlas, 185
   G-, 187
Attaching a cell, 35, 145–148, 153, 204, 233

B
Base
   change, 27
   of a filter, 68
   of a map, 27
   of a vector bundle, 187
Boundary, 11
   map, 220
   simplicial, 222
Braid group
   full, 170
   pure, 152
Branch of the logarithm, 159
Brouwer's fixed point theorem, 121
Bundle map, 193
   between principal bundles, 191

C
Cantor set, 67
Category, 14
   opposite, 15
   simplicial, 219
   small, 15
Cell attaching, 35, 145–148, 153, 204, 233
Chart, 184
   change of, 187
Clopen, 42
Closed
   map, 12
   subset, 10
Closure, 11
Clump topology, 8
Cofibration, 204
Compact, 55
   locally, 59
Compactification
   one-point-, 62
   Stone–Čech, 71
Compactly generated, 78
   locally-, 79
Complete, 63
Completely regular, 52
Composition, 14
Conjugation, 87
Connectable (points), 105
Connected, 41
   locally, 45
Continuous, 2, 9
   sequentially, 4
Contractible, 114
Convergence, 4, 9
   filter, 69
   pointwise, 5, 25
Coset, 90
Countability axiom
   first, 4
   second, 9
Cover, 55

Covering, 155
   branched, 161
   Galois, 177
   normal, 177
   pointed, 181
   regular, 177
   universal, 177

D
Deck transformation group, 160
Degeneracy maps, 220
Degenerate simplices, 222
De Morgan's laws, 10
Dense, 11
Discrete space, 8
Disjoint union, 29
Disk, 34

E
Embedding, 22
Equivalence
   of categories, 133
   natural, 132
   relation, 30
Equivariant, 88
Essentially surjective, 133
étale space, 211
Exponential law, 76, 81

F
Fibration, 199
Fibre, 64, 155
   product, 26
   sheaf, 210
   transport, 167
   typical, 184, 185
Fibre bundle, 187
   associated, 192
   G-, 192
Figure-eight knot, 22
Filter, 68
   base, 68
   convergence, 69
   neighbourhood, 68
   ultra-, 69
Fixed point property, 121
Fixed point space, 93
Fixed point theorem
   small, 125
Fool's cap, 146
Fully faithful, 133

Function
   Urysohn, 48
Functor, 106
   adjoint, 213
Fundamental group, 130
Fundamental groupoid, 128

G
Genus (of a surface), 150
Germ of a section, 210
Grassmann manifold, 92
Group
   automorphism group, 16
   free, 147
   symmetric, 95
   topological, 85
Groupoid, 129

H
Hausdorff space, 47
Homeomorphism, 15
Homogeneous space, 91
Homotopic, 111
Homotopy, 111
Homotopy equivalence, 114
   weak, 230
Homotopy equivalent, 114
Homotopy groups
   higher, 130
   stable, 205
Homotopy invariant, 116

I
Identification, 30
Immersion, 124
Interior, 11
Intermediate value theorem, 43, 52, 185
Interval, 42
Isomorphism, 15
   natural, 132

K
Klein bottle, 153, 157, 165, 167, 169, 173, 177,
   178, 182, 185, 188, 193
Knot, 22
   figure-eight, 22
   trefoil, 22
Kuratowski's closure axioms, 13

Index

L
Lebesgue number, 58
Lens space, 154, 195
Lift, 117
Linearisation, 234
Locally compact, 59
Locally-compactly generated, 79
Locally connected, 45
Logarithm, 160
    branch of, 159
Loop, 104
Loop space, 104
    free, 104

M
Map
    adjoint, 75
    boundary, 220
    closed, 12
    continuous, 2, 9
    degeneracy, 220
    equivariant, 88
    locally trivial, 185
    open, 12
    proper, 64
    separated, 53, 162
    simplicial, 222
    trivial, 184
Mapping class group, 109
Mapping degree, 119
Mapping torus, 38
Mayer–Vietoris problem, 109, 137, 147
Metric, 1
    discrete, 2
    induced, 2
    space, 1
Möbius strip, 32, 38, 64, 152, 161, 185, 188, 197
Monodromy, 167
Morphism, 14
    automorphism, 16
    endomorphism, 16
    identity, 14
    isomorphism, 15

N
Natural, 131, 132
Neighbourhood, 7
    filter, 68
    in metric spaces, 2
Normal, 47
Normaliser, 94
Null-homotopic, 113

O
Object, 14
    cosimplicial, 221
    simplicial, 221
    terminal, 220
Open
    map, 12
    subset, 7
Orbit, 88
Orbit category, 174
Order topology, 10

P
Partition, 30
Path, 101
Path component, 106
Path component functor, 107
Path-connected, 105
    locally, 110, 170
Path space, 101
Poincaré sphere, 180
Pointed
    covering, 181
    space, 181
Presheaf, 208
Pretzel surface, 149
Principal bundle, 189
Product
    external (of coverings), 158
    free, 138
    internal (of coverings), 158
    of simplicial sets, 224
    of spaces, 25
    symmetric, 95
Projection, 187
    canonical, 31
Projective plane, 150
Projective space, 30, 33, 53, 93, 153
Proper
    action, 96
    map, 64
Pseudocompact, 59
Pullback, 26
    of sheaves, 215
Pushout, 33, 137

Q
Quiver, 36
    realisation, 36
Quotient, 31

R

Rectangle, 24
Regular, 52
Representable (sheaf), 208
Representation of a group, 89
Restriction, 208
Retract, 16
Riemann surface, 161
Rotation number, 124

S

Section
  global, 215
  of a map, 185
  of a morphism, 16
  of a sheaf, 208
Semi-locally simply-connected, 172
Separation properties, 47
Sequence
  exact, 147, 196
Sequentially compact, 60
Sequentially continuous, 4
Set
  $\Pi(B)$-, 167
  Cantor, 67
  linearly ordered, 16
  partially ordered, 16
  quotient, 31
Sheaf, 208
  fibre, 210
  representable, 208
  skyscraper, 211
Sheafification, 214
Simplices, 222
Simplicial mapping set, 224
Simplicial sets, 222
Simply-connected, 135
Sine space, 45
Site, 207
Skyscraper sheaf, 211
Space
  compactly generated, 78
  discrete, 8
  Hausdorff, 47
  homogeneous, 91
  metric, 1
  metrisable, 9
  pointed, 181
  projective, 30
  topological, 7
  total, 187
Square
  co-cartesian, 137

Stabiliser, 88
Stalk, 210
Standard simplex, 225
Star-shaped, 115
Stiefel manifold, 92
Structure group, 187
Subspace
  collapsing, 34
  topology, 20
Sum
  connected, 149
  external (of coverings), 158
  internal (of coverings), 158
  of spaces, 29
Surgery, 36
Suspension, 204

T

Target, 88
Theorem
  Brouwer, 121
  Heine–Borel, 67
  intermediate value, 43, 52, 185
  Seifert–van Kampen, 139, 144
  Tietze–Urysohn, 49
  Tychonoff, 68
Topological
  group, 85
  space, 7
  structure, 7
Topology, 7
  clump, 8
  coarser, 8
  co-induced, 28
  compact-open, 72
  discrete, 8
  finer, 8
  indiscrete, 8
  induced, 20, 24
  product topology, 25
  subspace, 20
Torus, 32
Totally disconnected, 44
Total space, 187
Transfer, 158
Transformation
  natural, 131
Transition function, 187
Transport groupoid, 129
Trefoil knot, 22
Trivialisation, 155, 184

U
Ultrafilter, 69
Universal property, 21
    co-induced topology, 28
    identification, 32
    induced topology, 21, 24
    pullback, 26
    pushout, 34
    sum, 29
Urysohn function, 48

V
Vector bundle
    complex, 197
    flat, 198
    real, 197
    tautological, 197

W
Weyl group, 94, 175
Winding number, 124

The manufacturer's authorised representative in the EU is Springer Nature Customer Service Centre GmbH, Europaplatz 3, 69115 Heidelberg, Germany. If you have any concerns regarding our products, please contact ProductSafety@springernature.com

Printed and bound by CPI Group (UK) Ltd, Croydon, CR0 4YY
26/03/2026
02078943-0011